Anonymus

Southern Plants - Southern Homes

Evergreen Lodge materials

Anonymus

Southern Plants - Southern Homes
Evergreen Lodge materials

ISBN/EAN: 9783742804976

Manufactured in Europe, USA, Canada, Australia, Japa

Cover: Foto ©berggeist007 / pixelio.de

Manufactured and distributed by brebook publishing software
(www.brebook.com)

Anonymus

Southern Plants - Southern Homes

INTRODUCTORY.

Spring, 1891.

ONCE more old Earth has finished its mighty course around the
sun, and the race of 1891 has well begun. The bright Spring
days with their balmy atmosphere will soon be upon us, when
thousands of our customers who have bright and happy homes will
now be planning for their Spring task in Floriculture. And so,
friends, in anticipation of your recurring wants, we take pleasure
in presenting you our Spring Catalogue for 1891, and trust it may
prove of some interest to you.

Our readers will observe the very modest appearance of our Catalogue this Spring. We have abandoned the flashy lithographs that
decoy the inexperienced purchasers, and the hundreds of exagger-
ated illustrations that mislead the uninitiated in plant buying.
We believe the era of tawdry catalogues is on the wane, and that the
flower loving public want better plants, bulbs and trees, with practical hints on their cultivation and care, in lieu of the densely
begrimed pages of the numerous catalogues now sent out by many
of the florits and seedsmen in the United States.

While our catalogue does not abound in highly colored plates, the
change does not in the least imply a change in our policy. On the
contrary we have determined to make our catalogue instructive to
our customers by helping them to so cultivate their plants that
they may be successful in all their undertakings, and thereby continue their patronage with us. Whatever in our policy accrues to
the benefit of our customers is to our advantage, as our customers'
interests and ours are inseparable. The stock of plants which we
catalogue this Spring is fully up with the times, and the increasing
demand for the unexcelled collection we offer is the best criterion
that our patrons appreciate our efforts to send out good plants, and
do a square and honest business. We enumerate nothing but what

is good, and quote them so reasonable that all may buy. Our Chrysanthemums cannot be excelled in America or elsewhere. We have given much time and attention to this class of plants for the past few years, so that at present we have the most extensive collection of this now all popular plant for buyers to select from in United States, having devoted several large greenhouses especially to their culture. The quantity in which we grow them is in no way at the expense of quality. At the Piedmont Exposition Chrysanthemum Show, held in Atlanta, Ga., on October 28th, 29th and 30th, 1890, our blooms carried first honors over all competitors. Also at the great Chrysanthemum Show held in Cincinnati, Ohio, on November 11th, 12th and 13th, 1890, where competitors from East and West met in full force, we were fortunate enough to win over the exhibits of many veteran growers a valuable premium for the excellency of our Tennessee grown blooms, and bring Southward a much coveted trophy fairly won in a well fought contest. We touch "novelties" but slightly, as there is more opportunity to be humbugged by investing in them than anything we know of. There is enough of good well known varieties in every class of plants that are sufficiently beautiful to give all the enjoyment to be found in their particular class without paying exorbitant prices for untried plants. Our patrons must also be assured that all new varieties are given full trial by us, and should they prove of merit, we will in due time offer them for sale in our Catalogue. We guarantee satisfaction in every case, and invite where convenient our customers to pay us a visit and see our greenhouses and flowers. We have, owing to our large shipments over the lines of the Southern Express Company, secured a special rate of twenty-five per cent. less than their regular charges, which is of some advantage to our customers. The many flattering letters received by us would fill a book as large as this Catalogue, all of which we are most grateful for, and considered by us a high compliment to our business methods, and we here wish to return to all our heartfelt thanks for the words of encouragement as well as for the liberal petronage given us, and it is almost needless to add that it shall be our aim in the future as well as in the past to endeavor to please each one of our customers. The Southern people can never have a large and enterprising floral establishment if by their joint co-operation they do not help to make it so. It is needless to send to Northern florists

for what you can secure nearer home. Apart from this, however, we would be slow to solicit a single order from any individual whatever upon the grounds that it is our good fortune to live in the Sunny South. It is only because we can do as well for you and in most cases better, that we ask for a share of your patronage. With kind wishes for a happy New Year to all our customers and friends,
We remain, most truly,

J. J. CRUSMAN, Proprietor.   JAMES MORTON, Manager.

A Few Points--Read Before Ordering.

OUR TERMS.
Our terms are invariably CASH with the order.

ORDERS.
Remember all orders, large or small, are shipped in the order they are received.

VISITORS WELCOME.
Visitors are always welcome, as we have something of interest for them to see at all seasons of the year.

EXPRESS.
All orders for goods not stating the mode of transportation will be s.nt by express at the purchaser's expense.

SEEDS.
We have gone entirely out of the Seed business, and will in future devote our attention to the growth of plants exclusively.

REMITTANCES.
In remitting, send Money Order or Draft; if in currency, invariably register the letter, as we will not be responsible for remittances otherwise made.

PLEASE USE THE ENCLOSED ORDER SHEET.
If you have other matter to write, use separate paper or the opposite side of the sheet, but do not mix the order up with other matter.

PACKING BY MAIL.
We use a strong wooden box to pack in. Plants are all laid one way and securely fastened, thereby avoiding any crushing or mangling of leaves.

POSTAGE STAMPS.
When you cannot procure a money order and cannot make change otherwise, we will accept Postage Stamps. We would prefer two, five and ten cent stamps.

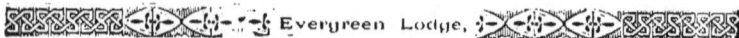

OUR AIM.

Our desire is to deal with our customers, that they may continue to favor us with their orders in the future, and they may rely on our endeavors to give satisfaction in every instance.

POSTAL NOTES.

Postal Notes are very convenient for small amounts, but are no more safe than money sent in an ordinary letter, as they cannot be duplicated if lost, and anybody that gets them can collect them.

HYACINTHS, TULIPS AND FALL BULBS.

We will issue aoout September 1st, a Catalogue of Bulbs and Winter flowering plants, which will be mailed to all of our customers who have ordered of us during the last year. Others who wish it are requested to write for it.

OUR SHIPPING FACILITIES.

Our shipping facilities are first-class, and the rates by express or freight from this point are much less to all the States South of us than they are from any of the Eastern floral establishments; the time occupied in transit is also much less.

HAVE YOUR GOODS SENT BY EXPRESS.

If possible, have your goods sent by express, is always our advice, as you invariably get your plants in better condition; and, as an inducement, we send a lot of plants extra to help pay express charges, and we feel ourselves repaid by the better satisfaction our shipments gill give you.

ORDERS FOR LESS THAN ONE DOLLAR.

No orders for less than one dollar will be filled unless fifteen cents additional to the price of the plants be sent to pay postage. It is quite as much trouble to handle, and requires nearly as much postage to mail, a fifty cent order as it does one for two or three dollars' worth of plants.

COLLECTIONS.

Parties having no knowledge of the different habits and adaptability of plants will do well to write and give us the particulars of the kind of beds they wish them for, or if for window culture; we can select them in most all cases to give the best satisfaction desired.

WHEN OUR LIABILITY FOR LOSS CEASES.

We take a receipt for all plants delivered in good order to carrying company, when our liability ceases, and the plants are at the risk of parties ordering. We make no charge for drayage or packing. If mistakes occur, notify us at once; otherwise, we are not responsible.

LETTERS AND PACKAGES.

Letters travel somewhat faster in mails than packages, so, if we write you a letter, and it reaches you before the plants, wait a day or two before writing, and give them the necessary time, and in ninety-nine

in every hundred, all will come out right, saving both of us the trouble of writing.

BE PATIENT.

In our busy season the office work is so pressing that packages of plants frequently leave the greenhouse before we get an opportunity to write, and as this is unavoidable, we beg our customers, if any plants are missing, to kindly wait two or three days for a letter of explanation before informing us of the shortage.

OUR LOW PRICES.

A careful comparison of our prices with those of other growers will show that we offer plants much cheaper than the same grade of stock can be purchased for at any other establishment in the country. In proof of this we are willing to duplicate the prices of any first-class and reliable house in the United States on plants, in this Catalogue.

C. O. D. ORDERS.

C. O. D. orders must be accompanied by at least one-fourth of the amount in cash, and the parties ordering are to pay the express charges for collecting. Large orders of shrubbery, trees, etc., can go by ordinary freight, by consigning to our own ordes and sending bill of lading by express, C. O. D., endorsed to parties ordering. Heavy express charges are thus avoided and collections facilitated.

OUR PLANTS.

Plants offered in this Catalogue are none of your puny, weak little things that take till Fall to see what you get; but are grown in pots ranging in size from two and one-half to four inches, according to the varieties of plants. We need not mention this to old customers who know what kind of plants we send out, but this Catalogue will be received by many people who never saw our plants or received our Catalogue before. Patrons can always depend on getting their money's worth, and more, too, as we are very liberal with our extras.

MISTAKES.

If anything is wrong with your order, do not think we intended it, for we have no interest in so doing; our interest is to give satisfaction, and that we are determined to do; so should an error occur, kindly put it down as a mistake, notify us, and we will put it right. We wish our customers would, in every case, keep a copy of their order, and verify it on arrival of plants; this will prevent mistakes as to what they "thought" they ordered, but which was never written upon their order sheet; and if not too much trouble, please drop us a card on the arrival of the goods. We are glad to know when you are pleased, and we wish to know of any dissatisfaction, that we may make it right.

ADDRESSES.

Please be careful and write your Name and Address plainly. We can readily make out what is wanted in an order, as we are acquainted with the names of all our plants, but we have no means of knowing what your name is or how it should be spelled unless you write it plainly. We receive many orders that are well written throughout, but when we look to see who sends them, the name is so carelessly written that we are obliged to guess at it. Indeed, some forget entirely to sign their name. Again we would say, "please sign your name carefully," as it will save us much annoyance, and possibly, prevent errors.

BULBS GIVEN AWAY.

Ten Handsome Presents to be Distributed.

AS an inducement to ladies to get up clubs, we have decided to offer as premiums a choice assortment of Spring and Winter flowering bulbs in ten separate collections, to be selected from our Fall Catalogue as soon as it is issued in September. The bulbs will all be of the finest quality, and will consist chiefly of Hyacinths, Tulips, Crocus, Narcissus, Lilies, Iris, Crown Imperials, and other popular flowering bulbs, to the amount of

FIFTY DOLLARS.

These presents are altogether an outside matter, and will have nothing to do with the other inducements offered elsewhere to originators of clubs. The parties that will secure the above presents will be notified by letter from us in May, when our shipping season closes, and their names published in our Fall Catalogue. We offered this inducement for the second time last Spring with good results. The names of the successful parties was published in our Fall Catalogue, and the bulbs sent to them as we promised. We will mail a Fall Catalogue to any one who may desire it. The bulbs will be distributed to the successful ladies or others as follows:

For the originator of the largest club, $20.00 in bulbs.
For the originator of the second largest club, $12.00 in bulbs.
For the originator of the third largest club, $5.00 in bulbs.
For the originator of the fourth largest club, $4.00 in bulbs.
For the originator of the fifth largest club, $3.00 in bulbs.
For the originator of the sixth largest club, $2.00 in bulbs.
For the originator of the seventh largest club, $1.00 in bulbs.
For the originator of the eighth largest club, $1.00 in bulbs.
For the originator of the ninth largest club, $1.00 in bulbs.
For the originator of the tenth largest club, $1.00 in bulbs.

Inducements to Clubs.

.·.

ALTHOUGH prices are low in this Catalogue for all classes of plants, most liberal terms are offered to friends who are inclined to obtain orders from others, and to secure thereby some fine specimens free of cost for themselves. In making up a club order it is important to state the sum sent by each member and the plants wanted, that they may be separately packed and confusion avoided when the plants are distributed. The full address of each is always required. The following are the rates from $3.00 to $20.00 (larger sums in proportion):

For a $3.00 club order the originator may select in plants 50 cents.
For a $4.00 club order the originator may select in plants 75 cents.
For a $5.00 club order the originator may select in plants $1.00.
For a $6.00 club order the originator may select in plants $1.10.
For a $7.00 club order the originator may select in plants $1.25.
For a $8.00 club order the originator may select in plants $1.50.
For a $9.00 club order the originator may select in plants $1.75.
For a $10.00 club order the originator may select in plants $2.00.
For a $15.00 club order the originator may select in plants $3.00.
For a $20.00 club order the originator may select in plants $4.00.
For a $25.00 club order the originator may select in plants $5.00.

CHEAP LIST.

— ·◊·

THE following collections, to be sent by express only, are very desirable to those who want a nice flower bed and care nothing about having the names put on each plant, the doing of which during the busy season consumes valuable time. We desire it distinctly understood that the plants in these collections are just as good and desirable in every way, and probably would be better than the individually selected plants at more than double the price.

One Dollar Collections.

Owing to a large increase in our facilities for raising plants this last Summer, we are able to offer plants in the following collections at the exceedingly low rate of twenty-five plants for one dollar, by

express only, and no premiums with these collections. If wanted
by mail, add fifteen cents extra:

25 Hollyhocks.	25 Violets.	25 Nasturtiums.
25 Verbenas.	25 Heliotropes.	25 Ageratums.
25 Coleus.	25 Tuberoses.	25 Salvias.
25 Pansies.	25 Gladiolus.	25 Cannas.
25 Geraniums.	25 Chrysanthemums.	25 Mahernia.
25 Achyranthus.	25 Centaurea	25 Tulips.
25 Carnations.	25 Asters.	

If the parties ordering prefer, they may select five plants from
five of the different collections; thus, five Pansies, five Geraniums,
five Tuberoses, five Chrysanthemums and five Verbenas. Not less
than five plants from any one collection to make up the twenty-five
as offered for one dollar. The plants will be packed nicely in a
small basket, and sent by express. Six collections for five dollars.

Collection of Plants for Five Dollars.

BY EXPRESS ONLY.

The following collection, containing one hundred and fifty-six
plants, for five dollars, is the cheapest ever offered, and every one
of them are fine strong plants that will grow rapidly and make
beautiful any home and its surroundings for the Summer, and
many of them will stand the Winter and come again the following
season. Nobody should neglect to beautify the surroundings of
their home with an offer like this before them. The selection of all
the varieties must be left with us. We cannot hunt up named
varieties at this price. They will all be packed carefully in a light
box or basket, and sent anywhere by express for five dollars. One-
half the collection for three dollars. No premium with this collection:

6 Everblooming Roses.	6 Pansies.	4 Violets.
6 Chrysanthemums.	6 English Daisys.	4 Rose Geraniums.
6 Achyranthus.	6 Paris Daisys.	4 Lantanas.
6 Fuchsias.	6 Verbenas.	4 Abutilons.
6 Geraniums.	6 Asters.	4 Sweet Allysum.
6 Tuberoses.	6 Ageratums.	2 Evening Glories.
6 Carnations.	6 Coleus.	2 Plumbagos.
6 Heliotropes.	4 Jasmine.	2 Hibiscus.
6 Gladiolus.	4 Hollyhocks.	2 Dahlias.
6 Begonias.	4 Feverfews.	2 Callas.
6 Salvias.	4 Mahernia.	2 Lantanas.

Mixed Dollar Collections.

BY EXPRESS ONLY.

Cheapness and merit being considered, it will be seen that where persons are not well acquainted with different varieties, so as to enable them to make judicious selections, these sets offer great advantages, and parties will usually get as good selections as they could make themselves, our aim being as far as possible to satisfy every customer that favors us with an order. The choice of varieties in these sets must be left entirely to us, as we positively cannot afford, at the collection prices, to permit customers to name the plants. These collections are subject to the following conditions: That persons ordering are only to name the letter or numbers, designating the collection or collections wanted, as it takes too much valuable time to read long letters, giving detailed lists of plants in each collection desired. Simply the letter of the collection is all sufficient. Entirely our selection of varieties. All labeled. These collections are not entitled to premiums. ☞Give this mode of purchasing plants a trial. We are confident it will please you:

A—25 Althernantheas, all sorts.
B—25 Geraniums, double and single, all kinds.
C—25 Verbenas, all sorts; makes a very fine bed.
D—25 Begonias, flowering varieties.
E—25 Heliotropes, different shades.
F—25 Coleus; makes a beautiful foliage bed.
G—25 Salvias, numerous kinds.
H—25 Ageratums, an assortment.
I—25 Carnations, splendid kinds.
J—25 Basket and Vase Plants, in variety.
K—25 Fuchsias, double and single.
L—25 Chrysanthemums, Jap. and Pom.
M—25 Roses, Geraniums, Coleus and Achranthus.
N—25 Geraniums, Verbenas, Coleus and Heliotropes.

O—25 Coleus, Heliotropes, Ageratums, Feverfews and Lantanas.
P—25 Carnations, Geraniums, Fuchsias, Begonias and Coleus.
Q—25 Petunias, Verbenas, Heliotropes, Feverfews and Lantanas.
R—25 Roses, Geraniums and Verbenas, assorted.
S—25 Abutilons, Carnations, Coleus, Ageratums and Verbenas.
T—25 Carnations, Chrysanthemums, Achranthus, Ageratums and Coleus.
U—25 Miscellaneous Plants, all kinds.
V—25 Tuberoses, Gladiolus, Cannas and Caladiums.
W—25 Geraniums, Tuberoses, Pansies and Gladiolus.
X—25 Pansies, Heliotropes, Lantanas and Geraniums.

One Dollar Mail Collections.

Look carefully at this offer for one dollar. There are many bright and happy homes throughout the South where intelligence is supreme, and consequently good flowers appreciated, that are not fortunate to have an express office convenient to them. To place our flowers within the reach of such people, we have prepared the

9

following collections that we will send free, postpaid, through the mail, for one dollar, packed carefully in a nice wooden box. Any one of these collections will make a handsome bed, and nothing helps to make a home more cheerful than a neat flower garden, however small. If preferred, parties may select four plants from any five collections, and make up their twenty plants in that way. Any three collections for $2.50, or six for $5.00.

20 Geraniums.	20 Gladiolus.	20 Heliotropes.
20 Chrysanthemums.	20 Verbenas.	20 Coleus.
20 Carnations.	20 Asters.	20 Achyranthus.
20 Tuberoses.	20 Fuchsias.	20 Violets.
20 Pansies.	20 Salvias.	

Mixed Mail Collections for One Dollar.

PLEASE ORDER BY NUMBER.

1—4 Geraniums, 4 Roses, 3 Coleus, 3 Heliotropes, 3 Pansies, 3 Verbenas.

2—4 Fuchsias, 4 Carnations, 2 Plumbagos, 4 Salvia, 3 Tuberoses, 3 Gladiolus.

3—6 Violets, 6 Daisys, 4 Pansies, 2 Hollyhocks, 2 Abutilons.

4—4 Lantanas, 4 Petunias, 2 Crape Myrtle, 3 Begonais, 4 Smilax, 3 Chrysanthemums.

5—4 Roses, 4 Carnations, 4 Hollyhocks, 3 Scented Geraniums, 3 Tuberoses.

6—6 Carnations, 6 Pansies, 6 Asters, 1 Calla, 1 Lily of the Valley.

7—6 Chrysanthemums, 2 Violets, 2 Roses, 2 Abutilons, 2 Salvias, 6 Fuchsias.

8—2 Crape Myrtle, 2 Feverfew, 4 Roses, 4 Hollyhocks, 2 Violets, 6 Garden Pinks.

There is not a home in the South or a person that receives this Catalogue, but what can use at least one of these collections to advantage. If you are boarding at a hotel and have no place to put them out, they will make a nice present for a friend not so situated, and will afford pleasure and remembrance all the Summer long. We ask therefore as an acknowledgment that this Catalogue is appreciated, an order for at least one dollar's worth of flowers, so that your name may go permanently on our books as customers, and continue to receive our Catalogues.

By Mail or Express for One Dollar.

14 Roses.	12 Calla Lilies.
14 Hardy Flowering Plants.	12 Succulent Plants, Sedum, Echeveria. Cactus, etc.
6 Evening Glories or Moon Flowers.	
10 Bouvardias.	14 Mixed Abutilons and Lantanas.
14 Begonias.	5 Lantanas, 5 Dusty Miller, 12 Coleus, 1 Abutilon.
16 Mixed Evergreens and Hardy Flowering Shrubs.	1 Azalea, 1 Camelia, 2 Dracenas.

These last two collections by express only.

"THE QUEEN OF FLOWERS."

ROSES.

❖

THE Rose is the "Queen of Flowers." No garden, however small, is complete without Roses. There are no flowers grown that are more universally admired than the Rose, and their cultivation is yearly extending, as it becomes more generally known that they are so easily grown, and that they can be procured at so trifling an expense. All that is necessary is to plant them in a bed of deep, fresh, loamy soil, well enriched with thoroughly rotted manure, and they are as certain to do well as a bed of Geraniums.

PREPARATION OF THE GROUND.

Roses will grow in any fertile ground, but are much improved in bloom, fragrance and beauty by rich soil, liberal manuring, and good cultivation. The ground should be subsoiled and well spaded to the depth of a foot or more, and enriched by digging in a good coat of cow manure or any fertilizing material that may be convenient. Renew old beds by decayed sods taken from old pasture land.

PLANTING.

When the ground is thoroughly prepared, fine and in nice condition, put in the plant slightly deeper than it was before, spread the roots out evenly in their natural position, and cover them with fine earth, taking care to draw it closely around the stem, and pack firmly down with the hand. It is very important that the earth be tightly firmed down on the roots. Budded Roses should be planted three inches below the bud. Always choose the most favorable time for planting in your own locality. Roses can be planted as soon as convenient after the frost is over. Always select an open, sunny place, exposed to full light and air. Roses appear to the best advantage when planted in beds or masses.

WATERING.

If the ground is dry when planted, water thoroughly after planting, so as to soak the earth down below the roots, and, if hot or windy, it may be well to shade for a few days. After this not much

11

water is required unless the weather is unusually dry. Plants will not thrive if kept too wet and without drainage.

PRUNING.

In most seasons it is best to prune established plants of hardy kinds in February. Tender varieties, such as the Tea Roses and newly planted Roses, may be left till a month later. As a general rule close pruning produces quality, and long pruning quantity of bloom. Climbing, Weeping and Pillar Roses should not be cut back; but the tips of the shoots only should be taken off, and any weak or unripe shoots cut out altogether.

PROPAGATION IN THE SOUTH.

For this purpose ripened or hard wood may be selected and the operation performed at any time from October to January. The cuttings in this case are usually made larger, generally with three or four eyes, after the wood is ripened enough to show the development of the buds in the axils of the leaf. The method most successfully practiced is to place the cuttings in a cold frame, or in the open air as far South as Savannah, Ga., Louisiana, Florida, the lower points in Mississippi, Arkansas, Alabama, and Southern Texas. The long heated Summers raise the temperature in the sandy soils in these sections as high as the atmosphere in the Winter months, if not higher in fact, forming a sort of a natural hot bed. All then that is necessary is to make the cuttings as above described and make a trench deep enough to plant them, leaving only one or two eyes or buds above the ground. Care must be taken to force the cuttings well in with the feet, so as to exclude the air. The cuttings may be set in trenches about four inches apart, and about two feet between the lines. Cuttings of Roses planted in this way in these States in November or December, will form roots by February, and if left to grow where placed without being disturbed, will by the following September make growths from one to three feet in height, according to the variety used. In parts North of these sections previously named, the cold frame had better be used for the cuttings. They ought to be inserted from two to three inches apart each way in November and keep above freezing through the Winter. Those not having the convenience of a cold frame, can do equally as well with the protection of ordinary garden or hand glass, or even some old discarded window sash

could be found and temporarily fixed up for the occasion. For the cold frame propagation it is not necessary to use sand exclusively, a good light sandy soil will answer for this purpose. One watering when put in to settle the soil around the cuttings, is usually about all that is necessary until they begin to root in Spring. Thus treated they will be rooted by March, and should be potted up, or the strongest might be planted out in Spring at a favorable opportunity, when cloudy and wet, for as no ball of roots can be lifted with them, it would not be safe to transplant them during a dry time or later in Summer. If neither potted up or planted out in Spring, they should stand unmolested until Fall, when they may all be planted out in the garden where desired.

INSECTS AND DISEASES.

As beautiful as the Rose is, and as gaily as she flaunts her regal petals to the envy of all other garden flowers, she too has her moments of disease, and a number of troublesome pests assail her. But he who loves his Roses will not allow any of these difficulties to impede the progress of the culture of his favorites, but rather he is incited to succeed in spite of these drawbacks. Knowing that a faint heart never won a fair lady, he cannot expect the smiles of Catherine Mermet or Marie Van Houte unless he thoroughly cultivates the acquaintance of these beauties and waits upon them with more attention and deeper concern than would the gallant of the ball room upon the attendant belles. The following are the chief foes with which the Rose has to contend:

THE APHIS

or Green Fly is well known to all who have grown Roses. It is a small green louse about one-eighth of an inch in length when fully grown. Through their slender beaks they suck the juices of the plant, always working at the tender shoots, and in a short time will, if unmolested, destroy the vigor or vitality of any Rose they infest. The best destructive agent to use against them is tobacco; if growing in a pit or greenhouse it may be burned so as to make a smoke. Care must be taken not to smoke it too much; better light applications and repeat a couple of times until the fly is dead. In sections where tobacco is plenty, a sprinkling of the stems or refuse from tobacco stemmeries among the plants will keep them away. We always keep plenty of tobacco stems in all our Rose houses, either scattered among the plants or on the floor or under the benches, and in consequence never have any trouble from Green Fly. If the plants are grown out of doors, and infested with fly, a liquid solution made from tobacco stems will be found an efficient method of working their destruction. Take some tobacco stems and place in a tub or vessel or some kind, and pour boiling water upon them until the liquid has the color of strong tea; after it cools off sufficiently to handle it, apply it to the Rose with a syringe or wisk broom; a little soft soap or whale oil soap added to the solution will greatly aid it in its efficacy.

MILDEW.

This is a fungous disease often caused by great and sudden atmospheric changes, and a long continuance of damp cloudy weather. The best remedy is sulphur, and should be applied the moment the disease makes its appearance, which is in the form of a white or grayish substance covering the leaves and causing them to crimple and become deformed. The plants should be sprinkled first with water so that the sulphur will stick; the best plan though is to apply it in the morning while the dew is upon the plants. After a few days the sulphur will all fall off and the mildew disappear. This treatment applies to Roses grown both in-doors and out, but if grown in a pit or greenhouse the best way is to mix the sulphur with water to the consistency of a good stiff paint, and apply it to the pipes or heating apparatus in the house with a brush. The fumes given off from this will at once check the ravages of the mildew.

RED SPIDER.

This is a most destructive little insect, and generally commits its ravages in a greenhouse or pit, and only make their appearance when favored by a hot and dry atmosphere. These are very small, scarcely distinguishable by the naked eye; if isolated, they are of a dark reddish brown color, found on the under side of the leaves, and cause the foliage to assume a yellow tinge, and soon make sickly the plants they infest. A few applications of whale soap dissolved in warm water, mixed with tobacco water, applied with a syringe and thrown upward so as to strike the underside of the leaves, will soon destroy them. This insect does not attack plants that are syringed with water daily, and all plants grown under glass, not in flower, should be sprinkled overhead with water daily.

BLACK SPOT.

This disease seems to be troublesome in many places, and Rose growers in the Northern States have suffered much from its ravages. It has of late made its appearance in many places in the South, although at present it is not generally known. The Hybrid Perpetuals and the Hybrid Teas appear to suffer most from it. As its name implies, it is a black spot that comes upon the leaves of the Rose, and gives it a somewhat blighted appearance. As soon as the plant becomes infested with it, it loses all its vigor and will cease to make further growth. The real cause of Black Spot is at present a disputed question, some contending that it first affects the plants through the leaves; others again contend that it is caused by a loss of root action; but as far as we have been able to investigate, there are no two cases exactly alike, so that it is very

hard to determine what is the primary cause of trouble in both cases, and this disease is to-day less understood by the most practical men in the business than any other disease peculiar to the Rose. Since the causes from which it emanates is so badly understood, it is of course equally difficult to suggest a remedy. When grown in greenhouses, the best means of checking the disease we have found is a healthy, dry atmosphere at night and a free circulation of air during the day, with a little fire heat to counteract any cold draughts. Where Roses are infested with Black Spot in the open ground, the best remedy is to cut the plants back and remove all leaves infested; when it starts to grow again the chances are that the Black Spot will not appear.

ROSE HOPPER.

This is another troublesome pest with which the Rose is afflicted in the open ground. It is a small yellowish white insect about three-twentieths of an inch long, with transparent wings. Like the Red Spider they prey upon the leaves, working on the under side. They go in swarms, and are very destructive to the plant. As they jump and fly from one place to another, their destruction is less easy to accomplish than is the case with other enemies. Syringing the plants with pure water, so as to wet the under side of the leaf, and then dusting on powdered hellebore or tobacco dust, will destroy or disperse them.

ROSE CATERPILLAR.

These are the young moths or butterflies, varying from one-half to three-quarters of an inch in length. Some are green and yellow, others brown. They all envelop themselves in the leaves or burrow in the flower buds. Powdered hellebore will prevent in a large measure their moving over the plants, but the only method of killing them that is really effective is picking them off with the finger and thumb and tramping them under foot.

ROSE SLUG.

These slugs are the larva of a saw-fly, about the size of a common house-fly, which comes out of the ground during May and June. The female flies puncture the leaves in different places, depositing their eggs in each incision made; these eggs hatch in twelve or fifteen days after they are laid. The Slugs at once commence to eat the leaves, and soon make great inroads upon the foliage if not checked. They are about one-half an inch long when fully grown, of a green color, and feed upon the upper portion of the foliage. The best remedies are powdered white hellebore, or a solution of whale oil soap.

CLEANLINESS.

We have given a list of remedies for controlling the ravages of the various pests which worry the Rose, and would remind our readers that prevention is better than cure. Cleanliness, a pure atmosphere and the free use of water upon the foliage of our Roses in the morning and evening, will be conducive to their healthfulness. A watchful care with systematic attention to watering and syringing will keep away insect pests that otherwise would come to torment us. In suburban gardens, where water is conducted in pipes from the water works, it can readily be applied with fine effect with the hose, and a vigorous application of water is as hateful to the insects in question as it is to fighting cats.

Standard List of Roses.

These Roses are all grown in two and one-half inch pots, and are from four to eight inches in height; they are vigorous and thrifty, grown especially for our mailing trade. We would like to have our list of Roses carefully examined, as it is without doubt the finest in the country. State what varieties you have, if selection is left to us, and we will not send them.

Price 10 Cents Each, $1.00 Per Dozen, Purchaser's Selection of Varieties, by Mail or Express. Our Selection, all Named from this List, FOURTEEN FOR ONE DOL-LAR, by Mail or Express.

Aline Sisley. Violet rose; a fruity pleasant fragrance.

Arch Duke Charles. Brilliant crimson scarlet, shaded violet.

Antoine Verdier. Rich, dark, carmine pink, slightly shaded with white.

Adam. Bright flesh salmon rose, extra large size, double.

Adrienne Christopher. A lovely shade of apricot, citron and fawn.

Andre Schwartz. Beautiful crimson: a free flowering variety.

Aurora. Creamy white, shaded dark rose and very double.

Arch Duchess Isabella. White, shaded with rosy carmine.

Agrippina. Rich velvety crimson. Few Roses are so rich.

Apolline. A clear pink, dashed with rosy crimson.

Alba Rosea. A beautiful creamy white with rose colored centre.

Bella. Pure snow white, splendid long, pointed buds, tea scent.

Bon Silene. A dark crimson rose, often changing to crimson.

Bougere. A bronzed pink, tinged with lilac; large and full.

Bella Fleur de Anjou. Beautiful silver rose with pointed buds.

Beau Carmine. Fine carmine red, very rich, good size, double.

Baron Alexandre de Vrits. A delicate rose, highly perfumed.

Canary. Light canary yellow, beautiful buds and flowers.

Catherine Mermet. Its buds are inimitable, faultless in form, and charming in their every shade of color, from the purest silvery rose to the exquisite combining of yellow and rose, which illumes the base of the petals.

Cornelia Cook. The flowers are of the clearest, snowiest white, and arranged in the most faultless and symmetrical manner.

Cels-Multiflora. Full and double, pale flesh, deepening to rose.

Charles Rovolli. A lovely shade of brilliant carmine.

Clement Nabonnand. A coppery rose, tinged purplish crimson.

Countesse Riza du Parc. Coppery rose, tinged with soft violet.

Coquette de Lyon. A fine yellow rose, large, not at all formal.

Clara Sylvain. Creamy white, good, full form, and fragrant.

Comtesse de Barbantine. Flowers large, beautifully cupped, full and sweet.

Cloth of Gold, or Chromatella. Sulphur yellow of good substance and form; full and double; very sweet.

Crimson Bedder. Bright fiery velvety red; recommended.

Douglass. Dark cherry red, rich and velvety; large, full, double and fragrant.

Devoniensis. Magnolia Rose. Beautiful creamy white.

Duchess of Edinburgh. Buds the most intense deep crimson.

Estella Pardell. A strong grower; fine buds of the purest white, with light yellow centre; blooms in cluster.

Etoile de Lyon. Chrome yellow, deepening in the centre to pure golden yellow, flowers very large and double.

Hermosa. Light pink; good bloomer.

Isabella Sprunt. This Rose is a sport from Safrano, which it resembles in all respects save in color, which is a bright canary yellow.

Jean Pernet. A beautiful pale yellow, suffused with salmon; of medium size; beautiful buds.

Jean Ducher. Yellow shaded salmon; a strong and vigorous grower and profuse bloomer.

La Nankin. Apricot yellow; fragrant; good form; very distinct.

La Sylphide. Blush, with fawn centre; very large and double.

La Jauquille. A saffron yellow; distinct and always in bloom.

Lauretta. Blush white, with peach centre, sometimes dotted with pink; very double and sweet.

La Princess Vera. A creamy white, bordered with coppery yellow; very full and sweet; a good new rose.

Lamarque. White, with yellow centre; large, full flowered; very fragrant.

La Sylphide. Blush, with fawn center, very large and double.

La Tulip. Creamy white, tinged with carmine, full and fragrant.

Lady Warreuder. A pure white, sometimes shaded with rose.

L'Elegante. Vivid rose, yellow centre, shaded with white.

La Pactole. Elegant buds, color pale sulphur yellow.

Louis Phillipe. Rich dark velvety crimson, free and beautiful.

Lucullus. A beautiful dark crimson maroon, full and fragrant.

Louis Richard. A coppery rose, beautifully tinted with lilac.

Louis de la Rive. A flesh-white, inclining to a rose centre.

La Nuancee. Rose, tinged with fawn and coppery yellow.

La Chamoise. Nasturtium yellow; very beautiful buds.

Mad. Berard. Apricot yellow; occasionally golden yellow; large and double; of good substance and very sweet.

Mad. Celina Norey. A delicate shade of rose, the backs of petals purplish red; very large, full and of good habit.

Mad. La Countess de Casuerta. Coppery red flowers; large; petals of good substance, but not full; splendid buds for bouquets.

Mad. Meline Vellermoz. Creamy white, thick petals, large and full, and slightly fragrant; an excellent variety for planting out.

Mad. H. Jamin. White centre, shaded yellow; large and full.

Marie Ducher. Vigorous and tree growing; large, full flowers; color salmon a fawn centre; splendid variety.

Mad. Joseph Schwartz. White, beautifully flushed with pink, of good size; cupped, and borne in clusters.

Mad. Louis Henry. Flowers medium to large silvery white, shaded yellow; fragrant and of good form.

Moiret. Pink shaded salmon; very good Rose.

Mad. Bravy. Creamy white; large, full and very symmetrical.

Mad. Camille. A delicate Rosy flesh, changing to salmon rose.

Mad. Caroline Kuster. A bright lemon yellow and very large.

Mad. Chedane Guinoiseau. A beautiful yellow Rose with fine, long buds.

Mad. Margottin. A beautiful citron yellow, coppery centre.

Mad. Maurice Kuppenheim. Pale canary yellow, faintly tinged with pink.

Mad. Pauline Labonte. A salmon rose, large and full.

Mad. de Vatery. Red, shaded with salmon, of good form.

Marie Van Houtte. A lovely pale yellow color, with the outer petals most beautifully suffused with bright pink.

Mad. Hippolite Jamin. White, yellow centre, shaded pink.

Mad. Jure. Lilac rose, good size and substance, fragrant.

Mad'lle Rachel. A lovely Tea Rose, pure snow-white.

Marcelin Roda. Pale lemon yellow, with lovely buds and flowers.

Marie Sisley. A pale yellow, margined with bright rose.

Mad. Angele Jacquier. A light silvery rose, shaded yellow.

Mad. Falcot. Deep apricot yellow, with fine orange buds.

Mad. Dennis. Waxy white, centre fawn and flesh, large.

Mad. Dubroca. Delicate rose, shading to yellow.

Mad. Lambard. Rosy bronze, changing to crimson.

Mad. Welche. Pale yellow, sometimes cream, with short inner petals of glowing orange and copper.

Mad. Brest. Rosy red, shaded to crimson; large flowers.

Marie Guillot. Holds first place among white Tea Roses, in purity of color, in depth of petals and queenliness of shape.

Mad. Bosanquet. Flesh, shaded a deep rose, large size, sweet.

Mad. Damaizia. Salmon rose, changing to amaranth.

Mon. Furtado. Yellow, well formed, full and fragrant.

Mad. Jean Sisley. Pure white; an elegant Rose.

17

Marie Guillot. Holds first place among white Tea Roses, in purity of color, in depth of petals, and in queenliness of shape.

Marshal Robert. White, centre shaded with flesh; large, full and globular. 15 cents.

Marechal Niel. Bright golden yellow; very large, full and perfect form; good substance. This is unquestionably the finest yellow Rose grown.

Melville. Silvery pink; bright and elegant.

Niphetos. A large and very double white Rose of moderate growth; beautiful long pointed buds.

Perle des Jardins. Canary yellow; large, full, well formed; very fine. This undoubtedly is the best dwarf yellow Rose in cultivation.

Perle de Lyon. Yellow, with salmon centre; large, full and very fragrant.

Purple China. Rich, purplish crimson, velvety.

Pink Daily. Light pink flowers, produced in clusters.

Premium de Charrisseans. Bright carmine rose, fawn center.

Queen's Scarlet. Dazzling crimson scarlet, with beautiful buds.

Queen of Bourbons. A clear carmine, changing to clear rose.

Robusta. Clear carnation red, veined a rosy crimson.

Roi de Cramoisi. Bright purplish crimson, full and very double.

Regulus. Brilliant carmine, with purple and rose shading.

Rosa Nabonnand. Imbricated, delicate rose, vivid in centre.

Rubens. A creamy white, with flesh centre; very large and full; superb.

Veve d'Or. (Climbing Safrano.) Deep coppery yellow.

Safrano. Bright apricot yellow, changing to orange and fawn.

Solfatere. Sulphur yellow; large, double or full; fragrant; an excellent variety for the South.

Sombreull. Beautiful white, tinged with delicate rose.

Souv. de Mad. Pernet. Beautiful, soft, silvery rose.

Sulphureaux. Sulphur yellow, fragrant, fine in bud.

Souv. d'Elise Vardon. Creamy white, delicately shaded with pink; fragrant.

Souv. Isabelle Nabonnand. This is one of the most delicately colored Roses; has large globular buds of a charming light fawn and silvery pink.

Souv. de la Malmaison. A noble Rose; the flower is extremely large, quartered and double to the centre; color a flesh-white, clear and fresh; it has been considered the finest Rose of its class for thirty years.

Souv. d'un Amie. Rose, tinged with salmon; very large, full, highly perfumed; an old and reliable sort.

The Bride. The best pure ivory-white Tea Rose; the buds, which are of grand size, are carried high and erect on bright, smooth stems.

Triumph de Luxembourg. A rose carmine on a buff ground.

Vallee de Chamounix. The back of the petals are a bright yellow, the centre highly colored with glowing copper and rose, every tint being clear and bright.

White Bon Silene. This is a sport from the old Bon Silene, possessing the same vigorous growth and free blooming qualities, differing only in color, being a pure pearly white.

White Daily. Pure white, beautiful long pointed buds.

Wm. Allen Richardson. Orange yellow, centre a coppery yellow.

Hybrid Teas and Hybrid Perpetuals.

All these Roses are grown in the open ground with us here, and are strong and vigorous. We could catalogue several hundred varieties of these Roses if we thought proper to do so, the list of them is so lengthy, but to save our customers confusion we catalogue only a limited number, all of which are good, and give the finest variety as to color, fragrance and form.

Price 25 Cents Each, $2.50 Per Dozen.

Abel Carrier. A very dark crimson, with violet shade; centre a bright red; large, full and double.

Alfred de Rougemout. Pure white, double and lasting.

Annie de Diesbach. A bright rose, large and showy.

Caroline de Sausal. Pale flesh; large and full.

Beauty of Stapleford. A dark purplish crimson; flowers well formed and large; a very beautiful and distinct variety.

Coquette de Blanches. A pure white; large and full.

Duchess of Cannought. Most distinct in foliage and blooms; a delicate silver rose with bright salmon centre; large and highly scented.

18

Earl of Pembroke. Bright velvety crimson.

Empress of India. A dark violet crimson; double and fragrant; a splendid Rose.

Francois Levet. Clear bright rose; a fine grower.

Gen. Washington. Brilliant rosy crimson; a good bloomer.

Gen. Jacqnminot. Velvety scarlet, changing to crimson; a free bloomer; good for Winter forcing.

Hon. George Bancroft. Flowers large, full and regular; a bright rosy crimson, shaded with purple; very beautiful.

John Hopper. Bright carmine; good.

Jules Margotten. A bright cherry red; large, well formed, fragrant flowers; double and free; splendid sort.

La Reine. Clear bright rose; of fine form.

Lord Bacon. Deep dark crimson, shaded with scarlet.

Mad. Alexandre Bernaise. Salmon rose petals sometimes edged with blush; full and fragrant; a good variety.

Marie Bauman. Bright clear carmine of perfect form; very fine.

Merveille de Lyon. Extra large, blush white.

La France. A peach shaded rose, that blooms through the Winter season; double and fragrant; we consider this the finest Rose grown.

Nancy Lee. A satiny rose; a delicate and lovely shade; slender growth; flowers medium or small; very fragrant.

Oxonian. A beautiful lilac rose, shaded crimson; very large, double and sweet.

Prince Camile de Rohan. Very deep velvety crimson; large and full; a good Rose of splendid color.

Paul Neyron. Deep rose; very large and full; fragrant, free blooming; the largest variety known; very desirable.

Pride of Waltham. Delicate flesh color, richly shaded with bright rose; very distinct; flowers large and full.

Ulrich Bruner. Flowers large and full, with exceedingly large shell shaped petals; color cherry red; splendid variety.

Roses from Five Inch Pots.

We grow the following varieties in five inch pots. They are nice bushy plants, from twelve to eighteen inches high; were propagated last Spring and grown through the Summer and Fall in five inch pots.

Price 25 Cents Each; Our Selection, $2.50 Per Dozen.

Madame Etienne.
Madame Schwaller.
Countess de Frigneuse.
Charles Rovolli.
Louis Richard.
Homer.
M'dlle Cecile Brunner.
Sv. de la Malmaison.
Countess Laberth.
Adam.
Madame Dubroeca.
Madame Cusine.
Ulrich Brunner.
Mignonette.
Agrippina.
Marie Ducher.
Andre Swartz.
La Pactole.
Marechal Neil.
Paul Nabonnand.

La Janquille.
La Sylphide.
Perle des Jardins.
Aline Sisley.
Etoile de Lyon.
Bon Selene.
Hermosa.
Bongere.
Isabella Sprunt.
Madame Bravy.
Madame Brest.
Saffrano.
Catherine Mermet.
Sv. d'un Amie.
Croquette de Lyon.
Solfaterc.
Mad. Joe Swartz.
Cornelia Cook.
Mad. Damazine.
La France.

Duchess of Edinburgh.
Madame Celine Berthod.
Madame de Watteville.
Bourbon Queen.
Marie Van Houte.
Madame Camille.
American Beauty.
La Princess Vera.
Madame Lambard.
Marie Sisley.
Madame Margotten,
Louis de la Rive.
Queen's Scarlet.
Marie Guillott.
Sunset.
Pierre Guillot.
Madame Cusin.
Papa Gontier.
Niphetos,
The Bride.

Six Superb Roses.

The following six Roses are unexcelled in the hardy Remontan class; each one is a beauty. They are not new, but we have yet to find their equal, not to say superior, among all the Roses of recent introduction. They are large, strong plants, that will flower beautifully this Summer. This set gives all the range of color to be found in this class of Roses, from the purest white through all the shades of pink to the richest velvety crimson.

Price 50 Cents Each, the Set for $2.50.

Capt. Christy. Flowers perfectly formed, of large size, and very double; of a delicate flesh color, slightly shaded salmon; constant bloomer; a gem.

La France. Peach shaded rose, blooming through the whole season; double and fragrant; we consider this the finest Rose grown.

Paul Neyron. Deep rose, very large and full, fragrant and free blooming; the largest variety known, and most desirable.

General Jacqueminot. This is the best known and most popular Rose grown; it is perfectly hardy, free flowering, and very fragrant; color, dark brilliant crimson; this Rose should be most generally planted.

Baroness Rothschild. This might well be termed the "Queen of Roses," as nothing can compare with the massive beauty of its flowers, which are full five inches in diameter, and an exquisite shade of satiny pink in color; the heavy foliage comes close up to the flowers, making a most effective back-ground of green, thus giving the effect of a lovely bouquet rather than a single flower; it is perfectly hardy.

Mabel Morrison. A sport from Baroness Rothschild; flesh color changing to pure white; in the Autumn tinged with rose; double, cup shaped flowers very freely produced; it is the most beautiful Hybrid Perpetual grown.

Six Old Stand-Bys.

To the exclusion of many of the novelties now catalogued in glowing colors, we here offer our "six old stand-by" Roses, nothing new by any means, but good old Roses of the Tea and ever-blooming persuasion, with good clear records for growing vigorously, flowering abundantly, and pleasing everybody with their beautiful blossoms. We recommend these especially to anybody who has not already got them, or to parties not familiar with Roses and who want a good reliable selection to start with. They are large, strong plants, that will come into bloom soon after planting.

Price 50 Cents Each, the Set of Six for $2.50.

Marie Van Houtte. Yellowish white, the outer petals pink.

Sombreuil. Pale yellow.

Safrano. Apricot yellow.

Triomph de Luxembourg. Buff rose.

Devoniensis. Cream white.

Malmaison. Flesh, changing to pink; a gem.

Two Fine Roses for the South.

Last Spring we sold more Wm. Allen Richardson Roses than any other one variety we catalogued, except Marechal Niel. It is a beautiful Rose, and growing in favor more each day as it becomes known. It is a beautiful shade of yellow, of a deep bronze tint, a running Rose, and fine for porches or arbors. Belle Lyonaise is another fine companion Rose, a climber also, very vigorous, a pale yellow, fine strong wood, and handsome foliage. These two would make a handsome pair for a porch or door yard, or for greenhouse and pit culture.

Wm. Allen Richardson. Strong plants, 50 cents each.

Belle Lyonaise. Strong plants, 50 cents each.

New and Scarce Roses.

Nice Young Plants, Fifty Cents Each.

Clotilde Soupert. This beautiful variety is a cross between Polyantha Rose Mignonette and Tea Rose Mad. Damazin; the plant grows from sixteen to eighteen inches high, and is an excellent sort for either bedding or pot culture; the flower is large for this class, very full and finely imbricated; the outer petals are pearly white, shading to a fine rosy pink centre; very free flowering and beautifully scented.

Countess Anna Thun. A strong bushy grower, with flowers freely produced on short stems; flowers extra fine and large; color a rich orange yellow, shaded with silvery salmon.

Chateau des Bergeries. A large canary yellow bud, nearly equal to Perle des Jardens in size; recommended by its raisers as a Winter forcing variety.

Duchess de Bragrance. Light canary yellow; after the style of Coquette de Lyons, but stronger, and of better construction.

Dr. Grill. Medium size, vivid yellow, centre light orange, shaded pink; very exquisite fragrance.

Duchess of Albany. This variety is a sport from La France, but is for superior to it in every way, deeper in color, more expanded in form and larger in size; the flowers are deep, even pink, very large and full, highly perfumed, and of first quality in every respect.

Glorie de Margottin. Dazzling red, the most brilliant yet known; large, full and finely formed globular flowers; growth very vigorous, one of the most distinct hardy garden Roses in cultivation, and can be specially recommended for its vigor of growth, freedom of blooming and hardiness.

Luciole. Very bright carmine rose, tinted and shaded with saffron; base coppery; back of petals bronze; large, full and finely scented; good shape; long buds.

M'me Seipion Cochet. In bud yellowish pink, banded flesh rose, centre yellowish; a very free bloomer.

Meteor. A Hybrid Tea, of strong bushy growth, producing quantities of finely formed deep crimson-scarlet flowers; a very free and productive.

M'me Francisca Kreuger. Extra fine, orange yellow, tinted rose.

Mad. Etienne. Called in France Dwarf Mermet; it is a Rose of great promise; very free, producing its buds on short, stout stems.

Mad. de Watteville. This grand Rose is one of the most beautiful varieties lately introduced; the color is a remarkable shade of creamy yellow, very richly tinged with carmine, with large, shell-like petals.

M'me Hon. Defresne. A beautiful dark citron yellow, with coppery reflex; as charming in bud as in open flower; a strong grower and free bloomer.

Mad. Hoste. This is one of the finest Tea Roses that we have had for years, and is equally fine as a bedding Rose or for forcing for Winter flowers; it is a strong grower, and buds can always be cut with long stems; in cool weather it is an ivory, in Summer a bright canary with amber centre.

M'me Schwaller. This fine Rose has the strong, firm growth of the Hybrid Perpetuals, the same form and finish of flower, being especially beautiful when full grown; the color is a bright rosy flesh, paler at the base of the petals.

Princess Beatrice. A most beautiful Rose, and will be of value to all florists for Summer buds; splendid for pots and fine for bedding.

Princess Sagan. A strong growing Rose, with small closely set dark foliage, and medium-sized flowers of the brightest scarlet and velvety texture; a single bud or blossom will catch the eye at a great distance, so brilliant is the color; it is as free in bloom as Bon Silene, and almost unmatched in color, which fully atones for its lack of size.

Papa Goutier. Extra large, finely formed buds and flowers; full and fragrant and very beautiful; color a brilliant carmine, changing to pale rose; reverse of petals fine purplish red.

Souv. de Wooten. An American variety of great promise for forcing purposes, which is being largely planted by cut flower men; color rosy crimson, or crimson red.

The Queen. A most charming white sport from Souv. d'un Amie; will be useful for white flowers; we esteem this very highly.

Marechal Neil Roses.

We have undoubtedly the finest stock of this popular Rose in the country, both grafted and on their own roots. We have a continued demand for large grafted plants of Marechal Niel for planting in conservatories or greenhouses, or out-door planting in the extreme South. We have on hand now a large stock of these, grafted on stout stems, about five feet high, and can highly recommend them. Price, $1.50 each.

Fifty Cent Marechal Niel Roses.

We have a large stock of nice plants, very strong, that we will keep dormant until late in Spring, so they can be shipped safely and begin to grow and flower as soon as planted out. These always give satisfaction. Price, 50 cents each.

Four Excellent Roses.

PAPA GONTIER.

CATHERINE MERMET.

THE BRIDE.

PERLE DES JARDINS.

In our estimation these four Roses have no superiors as ever-bloomers. Perle des Jardins is the best yellow; Bride, the best white; Gontier, the best red, and Catherine Mermet, the best pink. Thus you have four distinct Roses, the best in their respective colors, that we can well recommend. Price for the four, one strong plant of each, $1.00.

Glorie de Dijon.

Next to the Marechal Niel as a most desirable Rose for the South, we commend the Glorie de Dijon. It can be grown as a running Rose for greenhouse or conservatory, or as a climber for the porch or fence, or may be cultivated as an ordinary bush Rose for the garden. In either case it does well, its large salmon colored blooms and fine vigorous foliage always making it a favorite. Price, strong plants, 50 cents each.

CHRYSANTHEMUMS.

THE Rose may be Queen of Flowers, but the Chrysanthemum is is the Queen of Autumn. Two sister sovereigns that reign simultaneously in their court. During the period that these floral monarch hold their levees each in turn is unimpeachable; both are lovely ever—their beauty is supreme. The smiles, the admiration and the praise of all are theirs. Dives and Lazerus avow their fond and complete affection for 'the twin monarchs in their respective courts. From early Spring until Fall the Queen of Flowers is monarch of her realm. The laudations of a chorus of ten thousand voices is hers. No rival in her court! no anarchy in her ranks! Serenely in regal pomp she sits in state where all may gather 'round but none dare mount the throne. Her monarchy is absolute—her crown is the people's smiles. But mankind is fickle ever, even with a monarch, for with the departing days of Summer the Queen that waxed in June wanes in November. Unhappy lies the head that wears a crown. What is to stay her tottering throne when the days of sere leaves are upon us? Coronets soon vanish at the people's will, and with the rustling of the November leaves, one by one her glories fade, and the magic sunlight of her petals are dim. Through the requiem of the Autumn winds we hear the music of Summer's last Rose, and with the ripening of the purple grapes her dynasty is deposed. The Autumn Queen is then by popular acclamation placed upon the throne, and wears her honors with an air befitting a stately queen that as she comes, is seen and conquers.

The Chrysanthemum is a flower with so many varieties in shape, such great diversity of color, and so differing in size; it comes at a season when all flowers are more appreciated than at any other time of the year; it is a flower requiring no expensive structure to bring it to perfection. The man who loves the Chrysanthemum, whether peer or peasant, has but little difficulty in getting good flowers. It does not depend upon the vagaries of fashion for its popularity, for it has intrinsic merit which has gained for it a position in America to-day such as no other flower, not even the Rose, ever had in the short space of a few years, although many others have shouldered for supremacy. In England our favorite flower is known by the soubriquet of "Mum," and the many associations devoted to its culture and advancement have their banners inscribed with the euphonious appellation of "Glorious Company of Mummers," which includes all who love this protean flower that gives its light to our lives when the sun of Summer has veiled itself from us, so that it may not appear to compete with the splendor and beauty of Mum. For it must reign alone, this peculiar flower, in the Sabbeth of the year, when nature has to change her garments and supply a special decoration for such a reception as Queen Mum is entitled to. However, although "Mum is the word," the injunction in this case must go unheeded, for where truth and justice do prevail, its peans must be sung and the timbrels of Gog and of Magog shall awake in its praise.

With us in America it is now popular beyond reckoning. No pampered inmate of the conservatory dare dispute its charms. From the East, and the West, and from the shores of Hudson Bay to the land of the Creole and palmetto comes the joyous general outcry of, "All hail the Autumn Queen." The ladies of the South who by their zeal and intelligent culture have for so long made their gardens bright with the Hyacinth and the Rose, and sweet with the Magnolia and Jassamine, have now fully taken up the culture of the Autumn Queen, and it is with pleasure we note its progress in their care. It proves as an agreeable companion in Fall as the Crocus, the Narcissus or Hyacinth in Spring. Spring flowers in a Southern garden come like a joyous prelude to a concert, but the Chrysanthemum like the closing strains of a parting song, whose fading notes linger through drear December, making that charming contradiction, a flower of Winter with the hues of Summer that tarrys with us until the Snowdrop ad Scilla rise to kiss their hands at parting. Chrysanthemums will then but blossom in the hearts of people who love them through the Spring and Summer months, but the glory which has been and the glory that shall be can never be erased from the tablets of memory.

It is not surprising to anybody who ever saw a large and beautiful display of Chrysanthemums to wonder for a moment at their increasing popularity. Some people unfortunately remember them before the skill of the hybridizer was brought to bear upon them, and picture them to themselves as the little pompons of dingy brown and yellow of old time gardens, forgetful that like Cinderella, once homely, they have been transformed into blooms defiantly beautiful, so that nothing in the whole range of floriculture is so truly captivating as a Chrysanthemum show. The glory of other plants, the perfume, seemed at one time beyond its reach, but patience has conquered even here, for the Progne, with the color of the amathyst, has also the odor of the Violet, while the chaste blooms of Nymphea, with its delicate perfume, resembles the oft sang of Pond Lilies. No further triumphs remain for it. Individually the Chrysanthemum may be said to be coarse when placed beside the Orchid, the Lily, or the Rose, but collectively all must bow before the Autumn Queen. What can equal a large display of well grown Chrysanthemums in November? Not all the Roses in Oakley or Forest Glen; not the finest array that slaughter or May could produce. The denizens of the Rose houses of Summit, of Madison, of Niles Center or Flushing must yield the palm to the Autumn Queen in November. The Camellias of Mobile and the Jassamins and Azaleas of Charleston, in which the mocking birds sing at eventide, though beautiful are not to be compared to the gay Chrysanthemum. The Orchids of Sebright, of Croning, of Breckenridge or Short Hills cannot reflect one iota of their beauty. The Dahlias of Saul, the Carnations of Swayne, or the Lilies of Pierson, pale into insignificance before a well arranged display of Chrysanthemums. The Summer glory of South Park, of Fairmount, of Alleghany or Central Parks, with all the beauty of their geometrical designs, with hundreds of thousands of bedding plants in all their prismatic, brilliant, flaunting colors, cannot equal the gorgeous hues or purity of the Chrysanthemum, or rival the splendor that gleams through their glossy foliage, of which the eye never wearies. The diversity is great, but the harmony is good. Each variety is queen in her sphere with no borrowed beauty. No chalk, no rouge, no pencilled eyebrows, no disposition but to please, and no aspirations but to grow and blossom in that location in which it has pleased its cultivator to place it. Such a flower as this has great claims upon our affection, and ere another decade shall mark its milestone to eternity the Chrysanthemum will be found in the hearts and homes of all people in America.

CULTURAL NOTES.

The many purposes for which the Chrysanthemum is grown requires each a special treatment, so as to obtain the best results in each particular style of culture ; thus, to grow large blooms for exhibition purposes, standards, bush plants for conservatory decoration, single stem plants for exhibition purposes, plants for ordinary garden decoration, for cut flowers for commercial purposes, single stem and single flower plants, all require a special cultivation that we have not the space to do justice to in this Catalogue. We, however, have pleasure in recommending to all lovers of the Chrysanthemum in the United States, "A BOOK ABOUT CHRYSANTHEMUMS" that contains all the information possible to accumulate regarding this now all popular plant. Is is written by Mr. Morton, the Manager of our establishment, and published by the Rural Publishing Company, of New York. It is rich, rare and racy, practical, historical and scientific, the first book ever written in the United States on the Chrysanthemum, and contains a vast fund of information for all its readers that we are confident will meet with a hearty reception from the thousands now devoted to the culture of the Autumn Queen. It is said that eternal vigilance inspired by reverent love is the price paid for good Roses, and he who would have beautiful Roses in his garden must have beautiful Roses in his heart. This trite but beautiful saying may with suitable appropriateness be transposed to suit the Queen Chrysanthemum. For it is this same vigilance and love, guided by an intimate acquaintance of its habits, its likes and dislikes, accompanied with an unyielding disposition to administer to its every want and care, and wait upon it at all times, and under all circumstances, through the storms of early Spring and Summer's drying sun and heat, that begets in Fall a glowing harvest of plants and blossoms that amply repays for all the attention given it. Providing at the same time that high cultivation always pays, and love's labor on the Autumn Queen is never lost.

TO GROW LARGE BLOOMS.

In view of the premiums we offer for large blooms in Fall, we give a few brief notes on this point, in the hope that it may be of interest to intending competitors. As to the varieties best suited to grow for this purpose, if asked to name twenty-five sorts that would produce the largest blooms, and to avoid as much as possible the high priced sorts, we would name as follows: Avalanche, Advance, Coronet, G. F. Moseman, Mrs. Irving Clark, Mrs. J. N. Gerard, Mrs. Gilmour, Lilian B. Bird, Harry E. Widner, Violet Rose, W. H. Lincoln, Madame Baco, Miss M. Wrightman, Mrs. Fottler, Grandiflorum, R. Bottomley, W. W. Coles, Molly Bawn, Mrs. W. Sargent, The Bride, Mrs. George Bullock, George Tyson, Mrs. F. Thompson, E. G. Hill and Etoile de Lyon.

SOIL.

The Chrysanthemum thrives best in a compost composed of three parts of fibrous loam, one part well rotted cow manure, with the addition of about a six inch pot full of bone meal to a bushel of this compost. A handful or two of soot added to this will also keep it free from worms and add to the vigor of the plant.

A FEW POINTS.

Secure nice healthy young plants in a fresh growing condition, choose a nice spot where they will have sunshine each day. Set the plants carefully out, taking care that the roots are moist and in no way suffering for want of water. As soon as the plants begin to grow place a stout stake to each one. The plants must not be pinched often and all regard for shape and formation of the plant must be lost sight of where large blooms only are desired. Consequently there will not be as many shoots as when pinched frequently, but they will be taller and less bushy. Some plants grow eight feet high, others not over half that height, according to the variety. When the very largest flowers are desired only one flower on each shoot should be allowed. All side shoots are

rubbed or pinched out from time to time, and the small flower buds are removed as soon as they are as large as raddish seed, which is during the latter part of August and September. The terminal bud is the largest and the one usually retained. Growers for exhibition often confine their plants to one stout stem, and every lateral shoot is removed as soon as they appear, and only one bud retained. By this means of devoting the complete energy of one plant to the development of a a single blossom, it is wonderful the size of the blossoms that are grown by this manner of cultivation, and where ample room can be obtained and a good supply of plants on hand, the intending exhibitor would do well to follow this rule if he wish to distance all competitors for large blooms in the November exhibitions. All through the season the plants must have constant attention. The watering and thinning of the shoots in Summer and the disbudding in September and October is the chief points in the raising of large blooms. The number of blooms allowed to remain on each plant is a matter that each grower must determine for himself, according to his circumstances and requirements, always remembering that quantity will always be at the expense of quality, as the flower producing power of each plant can be concentrated into one or more shoots, and the same power in each shoot can in return be concentrated into one particular bud. The number of shoots intended be be left on each plant must be decided on in the early Summer, not later than July 1st. Not more than six should be left, four would be better. Some growers for exhibition grow only one stem to the plant and one flower to the stem, but this is where neither labor or expense is a consideration. When the number of shoots are selected the chief care is to preserve only that many. All others should be rubbed off as quick as they appear and the selected shoots given all the chance possible to grow. Towards the first of September these shoots will begin to show their buds, when the process of disbudding must be closely attended to. This is a somewhat critical task, even to the experienced cultivator, as much judgment is required in the selection of the particular bud to be retained. Some growers take out the terminal or end buds as soon as they appear and secure fine blooms from some one of the side buds retained for that purpose. This is to be recommended in the case of early varieties that are apt to come in too soon, as the terminal bud will always blossom earliest. With all medium and late flowering varieties it is best always to retain the terminal bud. It is not safe as soon as disbudding commences to remove all buds but one. At the first disbudding leave say about three; this will be when buds are about the size of raddish seed. As soon as they attain the size of peas the third one must be removed. After they begin to develop rapidly and the fear of accident from grasshoppers past, the most shapely and promising bud must be retained, and the other removed. Keep a sharp look out for grasshoppers, as one of these buds just beginning to swell seems to be a favorite tit bit for them. Abundance of water must be given when buds have reached this stage, and liquid manure must be alternately applied and the foliage syringed each evening if the weather is warm with clear water. With close attention to these points there ought to be no trouble in producing some fine blooms. We hope in our Fall Catalogue to give full information about the cutting, preserving and shipping of blooms, which will be issued in ample time for all the information on this point to be seasonable.

INSECTS AND DISEASES.

MILDEW.

The Chrysanthemum being hardy and robust in constitution is singularly free from disease. When they are housed or sheltered in the Fall Mildew sometimes makes its appearance. This is caused by cold nights succeeding sunny days, or two great extremes of temperature. Over crowding the plants and insufficient ventilation is another fertile cause of Mildew. It must never be forgotten that it is shelter and not heat the Chrysanthemum wants, that is in sections where it does not stay out the entire Winter. Should Mildew actually appear, dusting the affected plants with powdered or flowers of sulphur is the best antidote, together with the maintenance of a dry atmosphere.

BLACK FLY.

The Black Fly or Aphis is the worst enemy the Chrysanthemum has to contend with; it infests the little plants in the early Spring, and will stick to them all the Summer long if not destroyed. This is closely allied to the Green Fly er Aphis that infests the Rose, and the same treatment will destroy it. A good decoction of tobacco water applied by means of a syringe or wisk broom, will make the plant so distasteful to them that they will soon forsake it. Soap suds from the laundry applied in the same way as the tobacco water, will also help to drive them away. In sections where tobacco stems can be procured cheaply, if they are kept sprinkled among the plants the insects will never molest them. If you have a water supply and a garden hose, and give your plants a good syringing each evening, it will also make it unpleasant for the bugs, and the chances are they will not want to stay on your plants any great length of time. If grown indoors, a smoking as recommended for Roses will kill them.

CATTERPILLAR.

In the Fall a sort of a brown Catterpillar preys upon them; there is no antidote for this better than the finger and thumb system of picking them off and destroying them.

Our Piedmont Prize Collection.

The collection of twenty-five plants grown from a single stem, and the one hundred cut blooms exhibited by us at the Piedmont Exposition in Atlanta, Ga., last Fall, was a revelation in Chrysan-

themum culture to the thousands who inspected them daily. The exclamations of astonishment and praise that we heard daily during the period of our most pleasant stay there would fill volumes, and was prized more highly by us than the money prizes we had won. We give below a list of the names of the varieties that composed the one hundred cut blooms with which we won the first prize, that we have designated our Piedmont collection. Price for the forty varieties, $5.00.

Advance.
Ada Spaulding.
Avalanche.
Alfred Werne.
Comt de Germiny.
Coronet.
Clara Rieman.
Domination.
Duchess Connanght.
Excellent.
G. F. Moseman.
Geo. Tyson.
Judge Reu.

John Thorpe.
Jossica.
Lividia.
Mrs. Irving Clark.
Mrs. J. N. Gerard.
Mrs. Alpheus Hardy.
Mrs. Gilmour.
Mrs. J. B. Wilson.
Mrs. Geo. Bullock.
Mrs. Wannamaker.
Miss M. Wheeler.
Mrs. A. Carnagie.
Moonflower.
Miss M. Wightman.

Tr. de la Exposition de.
Marsailles.
Prince Kamountski.
Royal Aquarium.
S. B. Dana.
Sachem.
Robt. Bottomley.
Gloria Rayonnante.
Molly Bawn.
Humboldt.
Martha Harding.
Violet Rose.
Mad. Baco.
Passemoy.

Regarding this display, we append a few press notices from entirely disinterested sources:

Atlanta Constitution: The Evergreen Lodge Flower Garden, of Clarksville, Tennessee, exhibited one hundred of the finest Chrysanthemum blooms ever seen in this city, and won the first prize with them from Mr. Woodruff, of Macon. The excellency of this display must be seen to fully appreciate the monstrous size and exquisite coloring of the blossoms.

Atlanta Journal: The first prize for the best twenty-five plants grown on single stems was won by Mr. James Morton, representing the Evergreen Lodge Flower Garden, of Clarksville, Tenn., who had on exhibition some excellent specimens of his skill as a plant grower. The first prize for one hundred cut blooms was won yesterday by this same well known establishment, the blue ribbon falling promiscuously over all their most interesting display.

Sunny South: The floral collection at the recent Piedmont Exposition from the nursery at Clarksville, Tenn., of Mr. J. J. Crusman, afforded a striking and refreshing spectacle and was the admiration of the thousands of visitors. It is pleasant to note that his rare pains and culture were rewarded by two substantial premiums and many testimonials of merit. For his exhibit of cut flowers and that Chrysanthemums he bore off a prize each of fifty dollars, while blue ribbons fell promiscuously to the remainder of his entries. Mr. James Morton, Manager of this famous nursery, presided over the excellent display, and by his polite attention to visitors and pleasant qualities generally, added much to the attractiveness of this department, and Mr. Morton became a prime favorite with all the ladies, many of whom were placed under obligations for his thoughtfulness. The Sunny South remembers him most kindly for special attentions, and many of his beautiful Chrysanthemums are now the wonder and admiration of all who visit our sanctum.

General Collection of Chrysanthemums.

The varieties enumerated in this list are all good. Among them will be found most of the novelties that were prize winners at the various shows last Fall. Price from this list 10 cents each, fifteen for $1.00, purchaser's selection; our selection, twenty for $1.00:

Aspasia. A large symmetrical flower with broad petals; outer florets soft pale satiny rose, centre purplish rose.

Alaska. A beautiful snow white.

Adrioudac. Round, full petals; white.

Albert Delaux. Very large fine flowers; petals large, incurved, silvery white; reverse tender rose.

Areguine. Good old bronze variety, of great merit.

Alba Plena. A compact double white; early and good.

Alleghana. Fine white large petals, resembling Carnation.

Agnes Hamilton. A beautiful rose and pink.

Bessie Fitcher. Deep rose pink with a lighter centre; a grand flower.

Ben d'Or. Handsome twisted yellow.

Bruce Findlay. Rich lemon yellow; a sport from Emily Dale.

Bould de Neigh. Fine white compact flowers; a very free bloomer.

Bicolor. Large flat flowers, red, striped orange; late.

Baron de Prally. Extra large flower; white, striped with lilac.

Bronze Jardin des Plantes. A bronze sport from Jardin des Plantes; fine show flower.

Belle Paule. Vigorous and free; white, shaded pink.

Brazen Shield. Bronze; late and of fine form.

Bronze Queen of England. A bronze sport from Queen of England; extra fine.

Boule Canaise. Dwarf; bright yellow.

Bettina. Beautiful clear bronze, incurved large flowers and long petals.

Belle Pottevine. Large spherical snow white, the most regular and perfect incurved.

C. Wagstaff. Pure white; the best of the Japanese type.

Capaucine. Centre a brownish yellow, incurved, a very large half globe.

Cythere. A bright rosy violet, immense flower in ball shape, and of even coloring.

Citron. Yellow, free and late.

Christmas Eve. A magnificent white of greatest beauty.

Christine. Peach, fine and distinct.

Comte de Germiny. Bright yellow, with broad petals, shaded bronze.

Charlotte de Montcabrier. Silvery white with silvery rose centre; long petals, tufted.

Callingfordii. A rich crimson, shaded scarlet; very large reflexed flowers, beautiful and distinct; this is the finest scarlet variety in existence.

Coquette. Petals twisted, gilded mahogany; yellow.

Diana. One of the best whites.

Duchess. Deep red, tipped yellow.

Duke of Berwick. Petals twisted, milky white, veined.

Duchess of Manchester. The best white Chinese variety in cultivation.

Domination. A grand variety; flowers a creamy white, large and fine; early and very handsome.

Duck of Teck. A rosy mauve, suffused white; distinct.

Elkshorn. An extra large and incurved flower of soft pearl color; after incurving towards the centre, the petals, which are nearly tubular, rise.

Etincelle. Red, shaded maroon, pointed golden yellow; flowers very large; a beautiful variety which is very highly esteemed.

Edouard Andigier. One of the finest of recent introduction; of enormous size; in color a rich velvety purple violet, with a silvery reverse; a shade scarcely to be found in this family.

Eleanor Bares. Rosy frosted white, centre cream, early.

Eugene Mezard. Amaranth; reverse of petals violet white, forming a ball at the centre; very fine.

E. S. Renwick. Silvery blush, reflexed; large size and fine.

Elaine. Pure white, broad petals, very fine.

Eva. Salmon, white and pink; flowers quite large; an acquisition.

Empress of India. A pure white sport from Queen of England; of same character.

Emily Dale. Rich primrose; flowers of large size and fine form; one of the best, incurved.

Exposition de Chalous. Pompone; free blooming white.

Frank Wilcox. Flower with very erect petals, slightly toothed, above medium size; rich golden amber, slightly shaded deep bronze; one of the best to last.

Fantasia. Cream white, pure and distinct.

Fair Maid of Guernsey. Flowers very large, of the snowiest white, in clusters; one of the best.

Fabias de Medina. Anemonie flowered; high centre.

F. L. Harris. Bright crimson red, a new and fine color; distinct and good.

F. T. McFadden. A reflexed Japanese with immense flowers, having broad, flat petals; the color is a rich mauve purple, an entirely new shade, and most desirable; will be a fine variety for specimens.

Fleur Parfaite. Rose tinted lilac; flowers large and early.

Frizou. Pure golden yellow; the flowers large, with centre petals whorled.

Gloriosum. Very light lemon, with immense flowers, and having narrow petals most gracefully curved and twisted; it well merits its name, and is one of the most attractive varieties we have ever grown.

Golden Bronze. Fine incurved bloom, vigorous habit.

Golden Rayonnante. A charming yellow variety of most beautiful color; the flowers are large, and borne in immense clusters; early.

Golden Queen. Fine deep yellow, large and late; a great bloomer.

Golden Dragon. Yellow, with twisted petals; one of the richest and handsomest colors.

Gorgeous. Golden yellow; a magnificent variety, early and distinct.

Grandiflorum. A magnificent variety; flowers of immense size, often six inches in diameter; petals very broad, incurving, so as to form a solid ball of the purest golden yellow; one of the very finest, and no collection will be complete without it.

George A. Backus. Ribbon like petals, of royal scarlet color; back of petals silvery rose; fine.

George Maclure. Nine inches in diameter; purple shaded amaranth; the outer petals tubular, inner ones broad, flat and incurved; promises to be the largest of its class; was awarded three certificates.

G. F. Moseman. An improved bi-color, being a brighter color; bright crimson, tipped with golden yellow; this has proved to be one of the finest of all the new ones.

Gertrude Henderson. Lemon yellow, with flat fringed petals; it lasts a long time.

Golden John Salter. A golden yellow, changing to amber; incurved.

Golden Queen of England. Very large and rich lemon yellow; one of the best; incurved.

Golden Fringe. Bright gold; most distinct.

Golden Empress of India. Same habit as Empress of India, but of a very handsome straw color.

Gold. As name denotes, this is of the clearest golden yellow, and is perfectly double; one of the best new yellows.

Gold Band. Handsome compact flower, round yellow petals.

H. Waterer. Late yellow, with reflexed petals.

Hiver Fleuri. Flowers large, fringed and of good size; very early and free; creamy white and blush.

Incomparable. Rich chrome yellow and old gold, mottled with bronze; a fine early variety.

Isle Japanese. Pink and rose, very free.

John Salter. Fine dark Chinese, round spherical blooms.

Jessica. Snowy white, with yellow centre; very large flowers.

Jennie Y. Murkland. Most distinct and very large, having a flat surface from which project long tubular petals, rich golden yellow, shaded with apricot and rose; a superb variety.

Josephine. Beautiful bronze yellow; free and good.

John Thorpe. Eight inches in diameter; full flowers with long broad petals, except the under row, which contains a few tubular ones; color the richest deep lake, a new shade; very vigorous and early.

Judge Rea. A delicate shade of pink; flowers seven and a half inches in diameter; a profuse bloomer, but its best feature is its earliness.

John Collins. Immense, large, and flat flowers of beautiful silvery bronze and ashes-of-rose color; very pretty.

John Welch. Dark crimson maroon.

John H. Bradbury. Deep crimson, tipped with yellow; fine reflexed flowers, in style like Duchess.

John M. Hughes. Awarded first prize and extra silver medal by Pennsylvania Horticultural Society in 1886; in color a beautiful silvery pink.

King of Crimsons. A good sized flower of the most intense rich crimson, and of handsome globular shape; this is an entirely new shade.

Leopard. The only spotted variety extant.

L. Canning. A most exquisite white, absolutely pure; the flower is quite regular in form, very large and flat petals.

Lambeth. Early, dwarf, and of a strong habit; purest white, and one of the largest early double varieties; the flowers are five to six inches across, borne in clusters.

La Dauphinois. Enormous flower, very double; yellow ochre, brightening as the flower opens; a very choice variety of good habit; first-class certificate.

Lord Wolseley. A grand variety; rich, deep bronzy red, shaded purple; one of the very finest.

Lady Slade. Soft pink, with lilac shade; most beautiful shape; incurved.

Louis Weille. Fine large flowers; violet mauve, lighter centre.

Lady St. Clair. One of the most beautiful of all whites; incurved.

La Desire. Fine white Pompone.

La Centaure. Light pink, changing to white; fine.

Le Tonkin. Centre white, shaded rose on the outside; flowers large, produced in bouquets.

Lady Talfourd. Delicate rose, silvery back, incurved.

Lord Byron. Large orange, tipped with red.

Lady Matheson. Large petals, reflexed at extremities; globular, and of a rosy cream color.

La Chinoise. Deep crimson, with finely twisted petals.

Lord Mayor. Carmine violet, with white ground, shaded rose; dwarf and free; a profuse and fine bloomer, opening last week in September; one of the best for pot culture.

Le Cygne. Flowers extremely large and of the most beautiful form; long tube shaped petals, creamy white; one of the best for all purposes.

Le Chevrefeullic. Flowers large, petals spiral; honeysuckle color.

La Triomphante. Pale lilac; reflex of the centre petals cream; very large and fine.

Lady Selborne. A very large pure white variety of the greatest merit and quite early in flowering; remarkable for its flakey and snowy-like whiteness.

Laciniata Rosea. A tender rose color, pointed gold; resembles very much the old Laciniata in effect.

L'Ebouriffee. Deep mastic yellow, with petals reflexed; a beautiful color.

Lorraine. Resembles Ben d'Or in shape, but of a brighter color; the flowers are borne on long, stiff stems; very full and fine.

La Nymphe. Delicate peach color, shaded white.

Louis Barthere. Very brightest crimson red, with under petals of old gold; the flowers are flat and very free; one of the finest of this much desired color.

Mad. Domage. Early white.

Medusa. Curiously twisted petals; weak grower.

Mrs. A. Blanc. Centre of floret erect; outer petals horizontal or drooping, of rosy lavender, centre soft clear rose, with a touch of gold in centre; an exquisite rosy flower and good grower.

Magnet. An immense drooping flower of reddish Heliotrope color; generally admired.

Mouaduoe. Tubular petalled yellow; of fine form.

Mrs. J. C. Price. An improved form of Golden Dragon; very handsome.

Mrs. Anthony Waterer. An immense spreading flower which has been produced eleven inches across, the individual petals being about one and a quarter inches; at first very pale blush on the outside, which tint it loses at maturity; when fully expanded, it has the appearance of a bunch of white ribbon.

M'me Drexel. A large Japanese variety in the general style of Mrs. F. Thompson; the flower is more incurved and the habit more compact of vigorous growth and a very free bloomer; outside of the petals silvery white, inside bright pink at the tips, shading to white at the centre; holds finely in bloom, and lasts well when cut.

Marsa. Rose, centre white; is very free, blooming in the form of a bouquet.

Mrs. R. Brett. An orange yellow; petals twisted, forming a perfect ball.

Mrs. Authony Wiegand. Beautiful soft pink; a decided improvement in this line of color; form excellent; producing flowers in great abundance.

Mrs. J. N. May. Exactly like Mrs. Frank Thompson, except in color, which is a soft clear yellow; these two are fine companion pieces, flowering at the same time.

Montplaisant. Petals incurved; crimson red at base, golden yellow at extremity; fine for bush plants.

Mrs. Richard Elliott. Another grand yellow, in every way distinct from Mrs. Price and Mrs. May; the form is regular, very double, showing no centre, very large, and slightly recurved; petals long, and of medium width; a grand exhibition variety.

Mrs. Howells. A very fine red and gold variety; outer petals broad and velvety and reflexed; the inner petals incurved, forming a round golden ball; very brilliant; a fine show variety.

Mrs. DeWitt Smith. Beautiful soft rose, changing to white at the centre.

Montgolifier. Dark amber, gold reflex.

Mrs. G. W. Childs. Bronze tipped, old gold, reverse shaded salmon; extra.

M. Norman Davis. Carmine rose, and dark lilac; early.

Mrs. R. S. Mason. Heavy petals, of cup form, of a light buff color.

M. Brunlees. Indian red, tipped with gold; incurved.

Mr. Gladstone. Deep chestnut red; incurved; fine shape.

Mand. Very bright pink flowers of medium size; flowers in clusters; very good for cut flowers.

Mrs. George Rundle. One of the most beautiful whites in cultivation; incurving, and a popular sort.

Mrs. Mary Morgan. Rich, deep pink; perfect shape; incurved.

Mrs. Littlejohn. Richest golden yellow, some flowers being marked with bright red; of medium size, prolific in bloom, and very effective.

Mabel Ward. Lemon yellow, shaded a silvery pink on back of petals; flowers very large and globular; a truly beautiful variety.

Mous. Roux. Red chestnut; like Baron Beust, only larger; incurved.

Mrs. George Bullock. A pearly white, flowers very large and flat; very fine for exhibition purposes.

Mrs. N. Halloek. Compact Chinese of a rosy pink shade.

Mad. Thibant. Fine dark red; valuable as a late variety.

M. Tariu. Silvery pink, very free and desirable.

Mrs. J. B. Wilsou. Resembling Mrs. F. Thompson; white and rose shading, changing to bright lavender; extra large blooms, and a first-class sort.

Mad. Bouebarlar. Rich mahogany; a vigorous and free bloomer.

32

Mrs. Wanamaker. Probably the most beautifully shaped pearly pink in cultivation.

Mrs. Frank Thomson. Large, incurved with broad petals; mottled deep pink, with silvery back, very distinct, flowers eight inches across.

M. Planchenan. Mauve, shaded rose and silver, flowers large, free and early.

M. A. Vilmorin. Medium size flower, full centre of beautifully whorled petals; reflexed petals of crimson and old gold, distinctly marked with crimson, beautifully twisted and undulating; very late.

M. E. Nichols. Salmon yellow, of medium size and in bouquets of four or five pretty flowers.

M. Norman Davis. Deep rose carmine, shaded lilac; large flowers, and one of the best.

M. Freeman. A grand flower, a silvery rose, shaded violet; very handsome.

M. Ghys. A very effective, much prized flower of satin pink, pointed with yellow; a fine variety.

M. John Laing. Richest crimson, of the largest size and very distinct and fine.

M. Boyer. Beautiful pink.

Mad. C. Audiguer. Flowers of the largest size, of the purest rosy pink; a gem.

Mad. de Sevin. Rosy amaranth, shaded with silver, a pleasing color; flowers very large and fine.

Md'lle M. Fabre. A silvery pink, with white shadings, large and finely shaped; beautiful.

Mad. Lacroix. Flowers of a light rose, changing into pure white; this is a superb variety.

Mary Salter. A beautiful feathery flower of creamy white; large and fine.

Martha Harding. Yellow, shaded with brown; large and full.

Minnie Miller. Dark rose, very free flowering; this may be described as the best rose colored.

Moonlight. Immense flowers of a pure white; a white Temple of Solomon.

Mr. T. Norris. Rich velvety amaranth; a reflexed flower of most brilliant color; golden centre.

Mr. W. Barr. Entirely distinct; base of petals the brightest crimson, partly tubular, with points of pure yellow; early, lasting a long time.

Mrs. Langtry. An enormous incurved Japanese; flowers one foot across, outer petals long and quilled, inside ones flat and most beautifully incurved; color pure white; charming.

Mrs. C. H. Wheeler. Flowers are of the largest size, and of such heavy substance that they appear as if stamped out of leather; color a bright crimson on the upper side of petals, while the under side is clear old gold, thus forming a most beautiful contrast.

Mrs. John Thorpe. A brilliant crimson; very decided in coloring; petals tubular for half their length, disposed in a very marked whorled shape.

Mrs. R. Brett. A distinct variety, differing from all others in its peculiar plume like flower and rich coloring of pure gold; a gem.

Mrs. E. G. Gilmore. Silvery pink petals, very large, partially quilted, incurving to the centre; this will make a fine exhibition variety.

Mrs. C. Carey. A magnificent variety, with very large, broad petals of pearly white, much curved and twisted; on first opening, the flowers show a disc, but afterwards the petals incurve and form a nearly perfect ball in shape.

Mrs. Wm. Meneke. Brightest yellow, with slender petals of a peculiar shape late.

Mrs. Cleveland. A pure white with long tubular petals.

Mrs. Vanaman. Cherry red; very large and perfectly distinct.

Naragansett. Reflexed clear white flower with white centre.

Nathan F. Reist. Flower snow white, of enormous size, broad petals and golden yellow centre; good for cut flowers, as it is fluffy in appearance and very pure in color.

Newport. The largest and best of its class a clear rose pink, opening flat and forming with age ribbon like balls of largest size; a splendid variety; first-class certificate.

Nellie Bly. Brilliant yellow, very large flowers, which are plumed and tasseled; a splendid variety.

Nuit d'Automane. The richest crimson amaranth, beautiful color; of the largest size.

Nut de Heiver. Old gold and bronze late.

Othello. Single, bronze red color, free and good.

October Beauty. One of the earliest varieties, flowering by October 1st; of medium size, good substance, lasting color, at first a dull pink, changing to white.

Osiris. Rose, shaded with lilac; of globular shape.

Old Gold. A free blooming sort of a handsome old gold color.

Phidias. Blush pink and silvery white; incurved.

President Hyde. Large, full double and reflexed twisted outer florets; color rich yellow; of fine habit and free flowering.

Prince Alfred. Rose crimson, shaded a silvery purple; incurved; very fine.

Perle Prectouse. Purple, tinted silvery rose; broad petals, and perfectly incurved high centre; a valuable addition.

Public Ledger. Named after the Ledger newspaper; softest pearly pink; a very large and so incurved as to resemble a ball.

Puritan. White, tinted with rose, large, good habit, and one of the finest for bush plants.

Prince of Orange. A brilliant yellow, shaded and edged with a narrow band of red; very fine.

Pelican. The finest of recent introduction; pure white shaded cream petals, irregular, flat, half tubular.

Phœbus. This is without any exception one of the finest yellow ever grown; the flowers are large and handsome; too much could not be said of this variety.

President Garfield. Brightest carmine: large flowers and very distinct.

Pietro Diaz. Brilliant red, of fine habit, large flowered.

President Arthur. Immense rose flowers, opening in whorls; was exhibited measuring seven inches across.

Purple King. Deep purple, rather late, but quite distinct in color.

Peter the Great. A most showy bright lemon yellow variety; beautiful foliage and fine habit.

President Spaulding. Very vigorous, of a rich reddish purple color.

Queen of Lilace. A handsome Chinese variety; lilac.

Queen of England. A very large, fine blush; partially incurved.

Robert Crawford, Jr. A seedling of Mrs Frank Thomson; color white, the under petals slightly tinged; very choice and large.

Rob Roy. Orange, turning to gold; yellow centre and globular form.

Robert Bottomly. Pure white, flowers large, very fine.

Rose Lace. Flowers medium size, each petal toothed; dark rose; very pretty.

Roseum Pictum . Very large, deep rose silvery reflex; fine habit and distinct.

Robert Cartledge. Fine yellow, broad incurved petals.

Ramona. Bright amber color.

Ralph Broklebank. A sport from Meg Merriluse; yellow, flat petals of irregular form.

Seuzon. Clear yellow, very brilliant.

Sabine. Soft canary yellow, of exactly the form of Timbale de Argent, that is a medium sized Anemone.

Snowdrift. Reflexed white flat petals, fimbricated; of medium size and robust habit.

Salterii. Brilliant red, reflexed; neat and very beautiful flowers, having many petals; deep golden yellow.

Sir R. Seymour. Deep bronzy red on upper petals, finely incurved; light rosy shade on the outside.

Soeur Melanie. Flowers small, reflexed of the snowiest white.

St. Patrick. Bronzy red, large, incurved and distinct.

Soleil Levant. Pale yellow, large quilled petals; a grand yellow.

Stars and Stripes. Streaked with pink and rose; free and good.

Sadie Martinot. A fine bright yellow: very late.

Souree d'Or. (Golden Stream.) Golden twisted florets, tipped yellowish brown: very large.

Syringa. Lilac; of immense size, centre petal increasing, other petals very irregular.

Thunberg. Flowers very large, petals long and much incurved; a pure primrose shade of yellow.

Thorpe, Jr. Fine yellow broad incurved petals.

Theodora. Rosy salmon with pale centre.

Troubador. Rosy pink; of fine form and large flowers.

Thomas Cartledge. A grand variety; pure orange color of great size.

Timbal d'Argent. Exquisite pure white Anemone flowered kind, the most admired of any in its class; blooms in the greatest profusion.

Tubiflorum. Tubular shaped petals, odd and interesting.

Talfour Salter. Rich, deep crimson, of large size, forming dense heads of flowers pointed with yellow.

Tragedie. Rather small, of a new shade of color, rose, pink and blush; neat and pretty.

Venus. Lilac peach; large and beautiful incurved.

Vallede Andore. Maroon, yellow shade, petals twisted.

Veil d'Or. Beautiful Japanese, yellow, broad petals; incurved and distinct.

Veir O'r. Similar to Yellow Dragon.

Walter W. Coles. Very bright reddish terra cotta, reverse pale yellow; outer petals long, broad, pointed and horizontal; centre short petals.

Webb's Queen. A late bloomer; a rich silvery rose; perfect bloomer.

Wm. Joyce. Single, rosy pink; very free and fine.

Winonah. This was shown as Blushing Beauty and indeed is a beauty; the base of the petals is pure white, laced with deep lavender pink; of the largest size, full and double.

Wm. M. Singerly. Rich plum purple, large and double; desirable for bush or standards.

W. A. Harris. Nankeen yellow balls, with red centre.

A Special Selection.

The varieties in this list are all good and desirable to grow either for exhibition blooms or specimen plants.

Price 15 Cents Each, $1.50 Per Dozen.

Alcyon. Very large carmine rose, striped white, reverse rose.

Carew Underwood. Bronze sport from Baronne de Prailly; first-class certificate

Advance. Incurved, of perfect shape, a deep pink, but quite distinct from every other kind; large flowers, double, and good for all purposes.

Alice Bird. A large, compact and well formed flower of intensely bright buttercup yellow, somewhat deeper in centre; one of the finest yellow varieties yet raised.

Brynwood. Very large flowers, a bright pink on inner sides of petals, outer surfaces clear silvery rose.

Baronald. A variety vieing with G. F. Moseman in beauty and size; flowers very large, of a rich deep red and golden bronze; very double and compact in shape.

Belle Hickey. Large, perfectly incurved flower of purest white, always finely incurved, and of good habit.

Dormillon. Rosy purple, large, straight florets of immense size.

Excellent. Very soft pink, somewhat resembling the color of the Mermet rose; flower very large, slightly drooping but quite double. Exhibited twelve inches in diameter.

Elanor Oakley. Clear, deep, bright yellow, large, of globular shape; of strong habit, and a grand specimen variety.

E. Molyneux. A rich deep maroon red, the outside, when expanded, being of richest golden tint; petals immense in both width and length. Four first-class certificates.

George Atkinson. Clear white, with flat petals, seven or eight rows deep; of immense size and strong habit, free flowering, and a variety useful for all purposes.

Golden Star. Richest clear yellow, of good size, perfect in shape, and very double; a grand variety for specimen plants; habit fine, and a free bloomer.

Governor of Guernsey. Golden yellow, similar in form to Peter the Great; fine late decorative variety.

Gladys Spaulding. Brassy yellow.

George Bullock. Rose pink.

Kioto. A beautiful incurved yellow of fine form and habit; no collection complete without it.

Lucrece. Pure white, resembling Christmas Eve, but surpassing that in size, form and lateness; largely used for cutting and late decorations.

Lillan B. Bird. Of the very largest size, with full, high centre; petals tubular, of varying lengths, the flower when fully open being an immense half globe; the color is an exquisite shade of "shrimp pink."

Lady Trevor Lawrence. An exquisite white, with broad incurved petals, a large flower and compact grower.

Mrs. Andrew Carnegie. Bright deep crimson, reverse of petals a shade lighter, broad, long and flat; of leathery texture, with strong, erect, heavy foot stalks; of robust habit, and a prize winner wherever exhibited.

M. Moussillac. Deep rich, fiery crimson, golden reverse; grand.

Mad. R. Owen. Extra large flower, composed of two or three rows of petals, most perfect snow white; large honey combed centre; one of the finest varieties of this section; curly.

Mrs. Alpheus Hardy. Flowers are pure white, medium size, incurved Japanese, the centre slightly indented, the disc entirely hidden; on the upper surface of the floret petals is what at first sight appears hoar frost or snow, which gives it a chaste, delicate and fluffy appearance. It is rather a delicate grower, and not adapted to out door cultivation, but succeeds best in a greenhouse.

Miss Esmeralda. Incurved, deep crimson, double flowers, well built, tips of petals having a decided silvery tinge, the lower row of petals being flat and a red coppery bronze. This is a grand and strong variety for decoration or exhibition.

Mrs. Irving Clark. Pearl white in the margin, shading to deep rose in centre, which is beautifully whorled; large size and fine.

M. Garnar. Bronze orange, changing to golden yellow, of fine size; first-class certificate.

Miss Meredith. Flowers very large, of a rich rose pink, incurved and distinct in shape.

Mrs. Jessie Barr. A fine incurved pure white of large size, with flat, ribbon-like petals.

Macbeth. Jonquil yellow, richly flamed dark crimson; a bright and striking variety.

Mrs. John Wanamaker. A superb variety, of perfect incurved form; blush lilac and silvery white; one of the best of recent introduction.

Mr. H. Cannell. Rich and broad petals, incurving in the most lovely shape, occasionally here and there a petal standing erect out through the incurve; color of brightest possible yellow.

M. Mathounet. A large, incurved, dark purple rose, lighter centre, almost changing to white with age.

Monadnock. Flower full, bright yellow tubular florets. Awarded silver medal by Massachusetts Horticultural Society.

Mrs. Fottler. Large, full, double flowers of clear, soft rose, the shade of La France rose; fine habit and free, and a splendid exhibition bloom.

M. Brunet. Lilac mauve, large, straight florets; fine show flower.

Mrs. M. J. Thomas. White, large size.

Mrs. E. W. Clarke. Deep purple amaranth, silvery rose reverse, very large and highly scented.

Mrs. S. Humphreys. Pure white, large.

Martha Harding. An English variety of great merit; old gold and bronze, long thread-like twisting petals, forming a close, solid mass, of largest size.

M. Pierre Destonbes. Petals three to four inches long, entirely tubular; soft rosy pink, tinged saffron at the centre; an odd and handsome curiosity.

Mrs. Potter. This is also known as Stonewall Jackson and Crystal Wave. Pure satiny white, beautifully undulating petals; late flowering, and good for cut flower work.

Mrs. Hugh Graham. Incurved petals of light pink, lined with white; the flowers large and full, of good habit.

Narragansett. Clear white, with a full, high centre.

Nymphæa. A new sweet scented variety, the flowers resembling a Water Lily, hence its name; has a most pleasing fragrance, is a vigorous grower, and a fine acquisition to any collection.

Prince Kamoutski. Large, incurved, of the Comte de Germiny type; inside of petals bright crimson, outer ones deep coppery bronze; very free and fine; a superb addition.

Rollin Thatcher. Flowers richest yellow; thick, solid and compact; petals forming a perfect ball. At in every respect for specimen plants.

Red Gauntlet. Dark bronze, a vigorous grower and free bloomer.

Shasta. A pure white, large spherical blooms, distinct.

Sachem. Reflexed strong growing clear yellow variety.

The Bride. This was disseminated two years ago without its intrinsic worth being known, but time has demonstrated its value, and it has been placed at the head of all white varieties.

T. C. Price. A perfectly double flower, much twisted and incurved in the form of a corkscrew. Strawberry cream color, very distinct, and thoroughly good for all purposes; very large.

Theodora. Salmon rose spiral, straw colored centre, large, fine, and the best for specimen plants.

Violet Rose. A grand double variety, exceedingly free, of perfect form, a beautiful combination of violet and rose in color, and for exhibition purposes all that can be desired.

Wm. H. Lincoln. A magnificent golden yellow variety, with straight, flat, spreading petals. An extra large flower, completely double, and of great substance.

A Choice Selection from All Sources.

The following list comprises the cream of all Chrysanthemums in cultivation at the present time. All the new varieties of last Spring that were sent out by both American and foreign raisers that have proved to be desirable are here enumerated.

Price 25 Cents Each, $2.50 Per Dozen; Our Selection, $2.00 Per Dozen.

Addie Decker. A new dazzling shade of mandarin yellow, enlivened with salmon and flame color. A showy variety, good for specimens.

Antoinette Martin. A glorious Japanese specimen of immense size; petals curling and intermingling, irregularly forming a compact mass of pink, silvery sheen.

Ada Spaulding. A striking variety of globular form, petals very large, broad and solid; color shading from the purest pearl white to a deep pink on the lower petals. A vigorous grower and an excellent variety for all purposes.

Cyclone. An enormous Japanese, with creamy white centre petals arranged in long whorls, forming a complete mountain like effect. This has been shown to measure eleven inches in diameter.

Crown Prince. A splendid improvement on Mrs. C. H. Wheeler, with broader petals and of a deeper hue of color, Ox-blood red on the upper surface, old gold beneath.

Clara Rieman. A rich lavender rose in color, shading to silvery rose, with white centre. A very large open surfaced flower of fine texture.

Carrie Denny. A clear amber, entirely distinct from anything in cultivation. A most novel and striking color. Comes in large spherical balls, incurving and slightly whorled.

Charles A. Reeser. A novel and peculiar shade of color, quite distinct, a violet pink, without shadings; a fine recurved variety of good habit. Splendid for pots, making a fine exhibition plant.

Claude Frollo. Large double pink variety of strong, robust, and perfect habit, a free bloomer, and an excellent variety for plant specimens.

Dango Zaka. Distinct and desirable variety, flowers of largest size. Color claret crimson and pink bronze, the centre incurving.

Eclipse. Large, flat, incurved, showing no eye. A bright mahogany color, and a showy and grand exhibition bloom.

E. H. Fitler. Flowers of immense size, incurved, with large broad shell-like petals. Color a mixture of gold, tawny yellow and bronze.

Eugene Giat. Dark red shading, lighter to centre, striped; bright orange, a valuable acquisition.

Fireball. Rich red, very high centre, the outer petals broad, flat, and short, sometimes streaked with bronze.

F. Marrouch. Bright yellow, wide, flat petals. Double, extra large and fine.

George Daniels. Enormous flower, nine to ten inches in diameter; petals very long and broad, measuring three-fourths inch in breadth; color silvery white, the reverse purple rose.

36

Edwin Lonsdale. One of the darkest varieties in cultivation; flowers of an immense size, resembling Mrs. Bullock in shape. Color deep cranberry, with a rich velvety appearance, that does not fade with age as some of the other dark varieties.

Garnet. A showy Japanese variety, the inner side of petals a rich wine red, the reverse silvery pink; on first opening petals have a peculiar manner of twisting or curling, showing the reverse color, when fully expanded they display the red shade.

Harriet Beecher Stowe. Pure white petals, long, flat and pointed, very free flowering; habit fine and compact. A good variety for bush plants.

Ivory. One of the finest whites in cultivation. Large, globular and early.

James R. Pitcher. Reflex flower of the Japanese type, very full and of great depth; color light delicate blush, turning to pure white as the flower matures. This is a strongly perfumed variety.

John Lane. A magnificent pink ball in appearance, fine for pots, splendid for cutting. Color a rose pink, with peach or light shadings on under side of petals, ends of centre petals tipped with gold. Flowers are borne on long, stiff, stout stems.

Mrs. J. J. Bailey. Winner of the Sunnyside Cup for the best seedling. Immense incurved flower, white with slight lemon tint in centre.

Mrs. Wm. Barr. Pure bright crimson, incurved so as to form a complete ball, under surface deep pink; large and of grand habit.

Miss Anna Hartshorn. A superb double variety, opens first pearl color, changes to white as it ages.

Mrs. S. Houston. A magnificent flower, large and double, pure white and of good habit.

Mrs. J. N. Gerard. A grand cup shaped variety, closely incurving with age. Of large size, and of the brightest peach pink.

Mrs. Gilmour. Silvery pink, very large and fine.

Mountain of Snow. A good white, very large.

M. A. Haggas. Fine incurved bloom of a light golden yellow. Sport from Mrs. Heale.

Mrs. Dunnett. An enormous flower of rosy blush color, points of petals quilled and prettily tipped with white, the same being very long and narrow. Good for exhibition.

Marie Ward. A grand and beautiful cup shaped variety, very double, and of large size; color purest snow white, the petals very long and somewhat narrow. It is a sport from Mrs. J. N. Gerard, with which it is identical, except in color, and is a fine exhibition variety.

Miss Alice Broome. Richest crimson, with yellow markings at base of petals and rich gold beneath; largest size and effective.

Mrs. Tyson. Large, full and double peach pink, of good habit.

Mons. Bernard. Bright violet amaranth. This is a grand specimen variety and the best of the shade.

Mrs. President Harrison. The largest of all the Mrs. Wheeler type, on which it is an improvement, both in constitution, size, color and habit.

Mrs. J. T. Emlen. Deep blood red on upper surface of the petals, under side old gold. A very large incurved flower of most splendid shape.

Mrs. Charles Dissel. An improved Mrs. Thomson, stronger than that well known variety, the flowers larger and perfectly incurved. Color variable, sometimes a soft lavender pink and at others cream white. One of the largest perfect-shaped varieties in cultivation.

Miss Minnie Wanamaker. Cream white, incurving from first opening to finish, when it appears as a snow white ball; is rather dwarf in habit.

Model. Large, full flower, of the deepest pink, similar to Grace Wilder Carnation. Of good habit, and suitable for all purposes.

Mrs. Lord. Fine clear yellow, medium size, a profuse bloomer, very double. A grand variety for pot work.

Mrs. L. P. Morton. Bright pink, base of each petal pure white.

Miss Mary Wheeler. Pearly white, of immense size, very double, the petals delicately tinted on edges pale pink.

Madame Baco. Extra large, deep rose, tipped golden, very double. A magnificent show bloom.

Mrs. Edmund Smith. A beautiful pure white, of an entirely new type of flower. Pure white, long narrow petals, of great substance and lasting quality, beautifully interlaced, being an entirely new type; an exquisite thing.

Mrs. Frank Clinton. Soft canary color, fading to straw color. Perfect in habit, extremely free, the flowers compact and slightly incurved.

Mrs. J. S. Fogg. Color a bright chrome yellow, long petals, large and attractive flower, of strong habit.

Mrs. W. K. Harris. One of the finest yellow varieties yet introduced, with immense double flowers thoroughly incurved and showing no centre, of a deep golden yellow; grand in size and form.

President Harrison. An immense cupped flower, the outside petals salmon red, centre deep Indian red. Enormous, free and distinct. Invaluable as a show bloom.

Peculiarity. An entirely novel flower, very double, one compact mass of bright rose tubes, having open mouth like extremities, of a bright rosy crimson, which are divided into from four to seven lobes; flowers incurve to centre, the florets when opening tipped with bright lemon or buff.

Passemoy. Copper yellow and butter color, large and full.

Reward. Reddish violet, of an immense size, large and spreading, distinct and free, and a fine exhibition variety.

Robert S. Brown. A magnificent dark crimson like Hon. John Welsh in color, but four times as large. Will make a magnificent exhibition variety, either as a cut flower or grown in pots. Color richest crimson, very bright and attractive.

Ramona. Incurved large full flower, the florets bright amber.

Robert Craig. A truly grand flower, very distinct, similar in shape to Mrs. G. W. Childs, but deeper in color and larger in size.

Rose Hill. Pale pink, about the shade of Grace Wilder Carnation; large size.

Royal Aquarium. A white, strongly-tinged muslin rose; the centre slightly cream.

Sunnyside. A delicate flesh-tint while opening, becoming white when fully expanded. Immense size and great substance.

Souv. d'Angele Amiel. White tinted carmine, long twisted, petals large.

S. B. Dana. A large variety, color dark orange brown, free flowering, vigorous and early.

Sunflower. The loveliest golden yellow Chrysanthemum ever raised; flowers extra large, and of brightest color, petals long and somewhat drooping.

Stanstead Surprise. This variety is well known and has taken the first rank as an exhibition variety. Rich rosy crimson, with silvery reverse, an immense bloom.

Tusaka Takaki. An immense spreading flower, exhibited ten inches in diameter, petals being three-quarters to an inch across. Blush, striped pink, each petal shading to straw color towards the centre.

Twilight. White showing at first, lemon centre. A beautiful large flower, invaluable for exhibition purposes.

Violet Tomlin. Bright purple violet on under side of petals, lined with light pink. A sport from Princess of Wales; very fine.

White Cap. Very distinct, close, compact flower, pure white, the under side of petals shaded violet pink.

Xantippa. A fine pure white variety, very high centre, almost forming a ball, the centre as well as guard petals being of the purest snow white.

Zago Tee. An immense spreading flat variety in style of Wm. Joice; petals long and broad, three or four rows deep, of bright dark pink.

Zenobia. A very large double Japanese, with long, flat, spreading and drooping petals of purest white; it grows to an immense size, and is a grand exhibition bloom.

A Few Superb Varieties.

Price 50 Cents Each, the Sixteen Varieties for $7.00.

Avalanche. Large, pure white, Japanese variety; full deep blooms, with long, straight florets; lower ones drooping when matured; dwarf and sturdy habit.

Coronet. Richest golden orange, incurving to centre; occasional crimson stripes on inner side of petals; size immense; a great improvement on E. H. Fitler.

Charity. Bright, rosy carmine, shading very light towards ends and centre of flower, which is six inches across; petals somewhat convex, and slightly incurving towards centre; outer row drooping.

Etoile de Lyon. One of the largest flowers in cultivation; a grand variety; deep lilac rose, shaded silver.

E. G. Hill. Immense bloom, of brightest golden yellow; full and very double; lower petals sometimes deeply shaded a bright carmine; an elegant variety of strong habit.

Eynsford White. As good and popular as Avalanche, and there has hardly been a first or second prize awarded but what this variety has been the best blooms in the stand; it has a broader and more ivory-white petal, and is the best and most solid white ever sent out.

G. P. Rawson. A superb double variety, very large, of an entirely new shade; a rich buff, with centre petals of bright nankeen and apricot yellow; is broad, nearly erect, and slightly whorling.

Harry E. Widener. Bright lemon yellow in color, without shadings; flower large, on stiff, stout stems that hold the flowers erect, without support; incurving, forming a large rounded surface; petals crisp and stiff, very free in their growth, but not coarse; this is the cut flower variety, and all that could be desired in the way of good color, fine form, and lasting qualities.

Mrs. Thomas A. Edison. A large and incurved flower with compact centre; one mass of long petals of the most delicate rose pink; very free, large, and so closely incurved as to resemble a solid ball.

Mrs. S. Coleman. Color a clear canary yellow, reverse of petals uniformly striped with rose and apricot shades, flowers large and deep; the grandest novelty of the strictly incurved type.

Mrs. Falconer Jameson. Large blooms of a chestnut bronze, tinted and striped with yellow, high centre and large; habit the best of its class.

Miss Mary Weightman. A magnificent and distinct chrome yellow bloom, of loose and feathery form, large and full flower, ten inches in diameter, early.

Mrs. Winthrop Sargeant. A brilliant straw color, incurved, flowers borne on long, stiff stems; very large, if not the largest in this line of color.

Rose Queen. A self color of bright, rose amaranth; very distinct shade; flowers broad cup shaped, five inches across; one row of petals quilled half way and then flattened on outer ends; centre well filled; profuse bloomer.

Molly Bawn. Those acquainted with Syringa will need no word of praise for its sport, Mollie Bawn; it is pure white, having been grown two years, which

shows its colors to be fixed; a most valuable variety, for its size, shape and purity.

V. H. Hallock. Rosy pearl, of a marked waxy texture; flower six inches in diameter; petals convex, rounded downwards half their length from centre, changing to a beautiful curved form; few outer petals, drooping towards the stem in a most graceful manner.

The Pink Ostrich Plume Chrysanthemum, "LOUIS BOEHMER."

This is one of the strongest growing Chrysanthemums we have. Its flowers are enormous, being nearly double the size of Mrs. Hardy, and it is absolutely free from any taint or blight, so that it is certain to succeed under the most ordinary care. It has the same wonderful hair-like growth or excrescences that appeared for the first time in the white variety, Mrs. Alpheus Hardy, but it differs from it in color, being a most beautiful shade of lavender pink, shaded with silvery pink on the ends of the petals. The flowers are splendidly incurved; the inside of the petals are deep rose, so that the contrast between their inner and outer surfaces is very decided, and adds greatly to the appearance of the flowers. Price 60 cents each.

New American Seedlings for 1891.

We have seen all of these in bloom at the different shows last Fall, and can recommend them as absolutely "first rate," with a record behind them that will warrant them premium winners at the next exhibitions. This set is par excellence. Single plants, $1.00 each, the set of ten for $8.00. Ready March 1st. Orders booked now, and filled in rotation.

C. W. DePauw. This was produced by the raiser of Widener; it is a very double sort, having long petals arranged in most perfect form; the color is a soft pearl-pink, with touches of light lavender; it has the appearance of a soft, fluffy pink ball of great size; will rank with any of the very finest varieties.

Elmer D. Smith. In the hands of our "crack growers" we believe that this variety can be made to outrank all others in the size of its diameter and the number of its petals; the foliage is very large and heavy, and dark green; the color is cardinal red, of a very rich and pleasing shade, faced upon the back of the petals with clear chamois; comes nearer being a scarlet maroon than any of the Wheeler type yet sent out; the flower attains a great size even under the most ordinary treatment.

Emily Dorner. This is a rather dwarf grower, but very sturdy; the flower is nicely incurved, petals broad, and of the richest shade of orange yellow, touched with crimson; received first-class certificate at Indianapolis, Ind., in 1889; color extremely rich; a Source d'Or in incurving form.

Flora Hill. What Widener is among yellows, Flora Hill is among whites; the finish of the flower is exquisite; it is of splendid size and heavy texture; outer petals horizontal or slightly recurving; the creamy centre is perfectly full and incurved; it is a good, clean grower with fine constitution; one of the most beautiful forms and most perfect whites in the whole Chrysanthemum family.

Frank Thompson. A splendid flower, very nearly spherical in form; petals very broad and heavy, and finely incurving; it is very nearly white in color, only showing a touch of pearl-pink at the base of the petals; a strong grower, carrying flowers on stiff stems; the blooms of this variety sold at $9.00 per dozen over the counter at Philadelphia last Fall.

John Goode. This is almost globular in form, of the finest silky finish, and destined to become a standard sort for cutting; the outer petals are of a delicate lavender, forming a decided band of color, the inner petals are clear lemon; a plant in bloom has a most beautiful airy appearance; it is a light, willowy grower, though strong and healthy; in all but color it resembles the popular variety, Mrs. George Bullock.

R. Maitre. Not surpassed by any pink in cultivation; of the largest size, perfectly double, and without a trace of coarseness; this variety will take rank with the finest; a splendid keeper, of most symmetrical form, and a thrifty grower; it perfects numerous flowers to the single plant; one of the six varieties which took the hundred dollar premium at Indianapolis; the color is delicate and exquisitely beautiful.

Philip Breitmeyer. A most distinct variety, having heavy stems and foliage of light yellowish green; the flower is of the brightest golden yellow, extremely double; petals rather short and of heavy texture; of Helianchus form; we predict that this will rank among the foremost as a pot variety, and also as a cut flower.

Mrs. I. D. Sailer. A flower of the largest size, finely incurving with broad, sharply pointed petals; a strong grower, producing heavy flower stems; the color is soft shell-pink, touched with lemon on the extreme tips of petals; its keeping qualities after cutting are extraordinary; as shown by W. K. Harris, it is one-third larger than Ada Spalding, and nearly a globe in form.

Sugar Loaf. It is the freest growing and freest flowering sort that we know, and under the most ordinary treatment produces quantities of flowers of the grandest size; the outer petals recurve slightly, while the inner rows incurve; the color is varying shades of yellow, often shaded bronze, sometimes perfectly clear; a giant grower, with corresponding constitution; this variety took the fifty dollar premium as the best seedling at the Cincinnati show, and was one of six to win the hundred dollars at Indianapolis. Extra large flowers can be cut with long, straight stems.

Dorner's Set for 1891.

Price one dollar each, the set of eight varieties for $6.50. Ready March 1st.

Mistletoe. Received first premium for best seedling, and each of the remaining seven were certificated at Indianapolis. Of the Comte de Germiny type, with the outside of the petals silvery white, lined within with crimson; has wide concave petals, incurving until nearly globular in form.

Mermaid. Very delicate yet bright pink, perfectly incurving broad petals, extremely delicate in color and finish; an extra good variety.

Mattie Bruce. Silvery pink in color and of medium size.

Anna Dorner. A full, fine bold flower, with outer petals striped and shaded a rich carmine, centre cream white.

Evaleen Stein. In the way of Bride, but an improvement on that variety, of a delicate white, with petals like Elkshorn.

Eda Prass. A fine, bold, and recurving flower of great substance and depth, a lovely white delicately shaded blush; of great promise.

Emma Dorner. A fine, deep violet pink, in the way of Violet Rose when finely done, but a deeper, purer color; large ball shaped flower of splendid substance.

Innocence. Seedling from Mrs. Hardy, and as fine in form, texture and finish as the parent, but without the velvety pile; the purest white found in the entire Chrysanthemum family.

Spaulding's Set for 1891.

Price one dollar each, the set of twelve varieties for $10.00.

Lily Bates. Very large, perfectly double, clear bright rich pink, petals broad and flat, a new and distinct form. Winner of the Pitcher Cup, certificate at Indianapolis, and medal at American Institute.

Clauncy Lloyd. A delicate flesh pink, changing to pure white; petals medium in width, flat and cup shaped, incurving and covering centre, outer row extending some distance beyond the ball form. Medal of excellence at American Institute.

Mrs. D. D. Farson. Unquestionably the most meritorious introduction of the year. Size immense, solid and compact; color bright Mermet pink. Winner of Pitcher Cup, silver medal of Pennsylvania Horticultural Society, medal of merit at American Institute, and certificates at Indianapolis and Springfield.

Mrs. Lay. A chaste and very beautiful large incurved flower. Petals are cup shaped, white with faintest blush lines on edges; extremely double, pyramidal rather than globular in shape.

Progression. Extra large late flowering variety, blooming about December 1st, and remaining in flower up to Christmas. Color purest white, very double, style, of Grandiflorum. Awarded medal of excellence at American Institute.

Charles Canfield. Extra large and full double incurved bloom of robust habit; color claret red, reverse silvery pink. Superior for exhibition.

Lizzie Cartledge. Very large, full and double flower, incurved except under row of florets, which reflex; color bright dark rose, reverse silvery white. It was exhibited at Philadelphia by Mr. W. K. Harris, where a plant carrying over two hundred blooms four to six inches in diameter was awarded first premium as one of the best six new varieties.

Mrs. R. J. Bayles. An immense incurving Japanese bloom, in style of E. H. Fitler and Coronet. Color clear yellow, striped and highly marked red, bronze and old gold; petals by actual measurement one and a half inches in width; of strong, robust habit, and the largest of all Chrysanthemums.

Mrs. Kendal. A fine Japanese flower with compact centre; color rich Jacqueminot, reverse of petals copper bronze, shading to gold from base to tips; a free bloomer of good habit.

Anna M. Weybrecht. Magnificent Chinese variety of purest snow white, the petals solid, broad and firm; of strong habit. Awarded silver medal by Pennsylvania Horticultural Society.

Mattie C. Stewart. Clear bright golden yellow, extra large and double; petals broad and flat, reflexing with age; high built bloom. Winner of silver cup at Indiapolis, certificetes by Pennsylvania and Hampden County Horticultural Societies, and medal of merit at American Institute.

John Smith. Novel and distinct shape. Extra double, a grand acquisition for show purposes. Petals cup shaped, and arranged in compact rows one above another, completely covering the centre. Deep Mermet pink, shading to silvery rose. Winner of Pitcher Cup, silver medal of Pennsylvania Horticultural Society, medal at American Institute, and certificate at Indianapolis.

Waterer's Set for 1891.
Price One Dollar Each.

Kate Rambo. Pure white, very large, broad double flower, florets curl at the tips and slightly incurve; fine, robust, yet compact grower, with strong flower stems; very distinct and a variety that is equally suitable for specimen blooms or growing as an exhibition variety.

Mrs. John Westcott. Color cream pink to cream white, of very pleasing shades; flowers enormous, reflexed, with very stout petals of most symetrical form, very stout flower stems and a grand and free grower; this variety has pleased all who have seen it, and has already made its mark as a cut flower variety, and is of just the style of flower that will be in demand for specimen blooms next season, and is equally available for exhibition purposes.

Mary Waterer. This variety opens up a new class in Chrysanthemums. In describing it we must make use of a new word, "recurved," for as Grandiflorum forms a ball by incurving, so does this form a perfect ball by recurving back to the stem. Flowers of a delicate rose shade, of immense size, very attractive, durable and very double; is of short but very healthy growth, and has good stiff stems, is exceedingly free flowering and make the handsomest of pot plants.

M. P. Mills. This variety stands alone as being the latest and perhaps the largest of all Chrysanthemums yet in cultivation. In shape it somewhat resembles a mushroom, and has very thick flowers of great substance; color an orange yellow, sometimes faintly streaked with red; its enormous flowers are well borne up upon the strongest of stems.

Mrs. Herbert A. Pennock. This is perhaps the greatest novelty of the times in new Chrysanthemums; is exactly similar in habit and shape to the well known Violet Rose, the flowers, however, are somewhat larger and of a beautiful orange yellow color; it has very strong stems, bearing its enormous flowers perfectly erect; it is valuable for specimen blooms, and is perhaps the strongest ever offered for that purpose.

Eldorado. A most lovely incurved deep yellow, possibly the most intense of all yellows, of dwarf, sturdy habit, with a strong flower stem, lasting a long time in flower, and an excellent shipper; it may be said this is the earliest and best early large incurved variety yet in cultivation, and a variety that will force itself into favor with all growers of specimen flowers, whether for exhibition or market purposes.

OUR PRIZES.
FORTY DOLLARS IN GOLD.

In order to further induce the higher cultivation of the Chrysanthemum in the South, we offered last Spring the above amount in prizes to the parties who would send us the largest cut blooms in

41

November, and there are few things we have ever done that has yielded us so much gratification. We have had competitors from every State in the South from Florida to Texas, and the blooms sent by all were truly magnificent, all raised from plants purchased from us, thus demonstrating beyond all doubt that our plants succeed admirably, and produce superb blooms in every section of the South. When we offered the prizes in Spring we did not expect to receive anything like the fine blooms all our competitors sent us, and each one has just cause to be proud of the excellency of their flowers, whether they were among the prize winners or not. The first prize of $25.00 was nobly won by Dr. D: T. Price, Booneville, Miss., who was only a few points ahead of Mr. James Balfour, of Gadsden, Ala., who gained our second prize. Mrs. T. A. Connor, of Cokesbury, South Carolina, won third prize with a superb collection. This lady must be highly complimented on the excellency of her blooms, and must remember that the third prize in a strong class of nearly two dozen competitors is an achievement to be proud of, beating numbers of the most prominent amateur florists in the South. We visited all the Chrysanthemum shows west of the Alleghenys, from Atlanta, Ga., to Chicago, Ill., and the blooms sent us raised from our plants grown in the South compared most favorably with those we saw in the great Northern shows grown by veteran cultivators. Upon this matter we congratulate our customers and ourselves. The prizes offered for next Fall will be as follows:

For the best 50 blooms, in 25 varieties (two blooms of each) first prize, $25.00.

For the second best 50 blooms, in 25 varieties (two blooms of each) second prize, $10.00.

For the third best 50 blooms, in 25 varieties (two blooms of each) third prize, $5.00.

These blooms must be sent to us by express any time from the 1st to the 15th of November next. There will be no entrance fee for this contest, but no one is eligible to compete whose name does not appear on our order book this Spring for a sum of not less than FIVE DOLLARS in Chrysanthemums. Any further information desired on this subject will cheerfully be given. In Fall Catalogue we will give full instructions as to how the flowers ought to be cut and shipped. The blooms eligible to compete for this must be of good size, only one bloom on each stem, and cut with a long stem.

AN UNPRECEDENTED OFFER.

To put the novelties of the present as well as the past season within the reach of all, we make the following unprecedented offer. Four sterling Chrysanthemums for one dollar. Send for these four and test them in your garden; you can afford it at this price. The following by mail or express for one dollar:

LOUIS BOEHMER. Pink ostrich plume.

MRS. A. HARDY. White ostrich plume.

ADA SPAULDING. Peach pink.

LILIAN B. BIRD. Tubular petalled pink.

☞These four rare varieties for one dollar.☜

CARNATION PINKS.

CARNATION PINKS, next to Roses, are the most popular flowers grown. Young plants should be procured in April or May, and be sure that they are young plants, no matter how insignificant they may look, for large plants are ones that have been bloomed all Winter, and are comparatively worthless. Carnations are quite hardy, and should be planted as early as possible, just as soon as the ground is in condition to work. The soil should be quite rich, well manured with thoroughly rotted manure, or, if not to be had, bone dust may be used to a good advantage. To have a beautiful bed of Pinks in the Fall, the plants should be set out about eight inches apart each way; as the plants grow, they should be "stopped," that is, when the shoots of growth become six inches long, they should have the points pinched out. The operation should be continued until the 1st of July, when it must be discontinued if flowers are wished in August. Price 10 cents each, fourteen for $1.00, purchaser's selection. Our selection, by mail, sixteen for $1.00; by express, eighteen for $1.00.

B. A. Elliott. This is the largest flowered Carnation in cultivation; a vermillion scarlet.

Brussels. Cherry red, with broad stripes of carmine.

Clara Morris. Model flowers of good size; very pure white, with the edges of petals marked with crimson.

Charles J. Clarke. A grand Carnation; pure white, fringed edge.

Buttercup. Magnificent yellow.

Dawn. Flowers very fragrant, color a blending of pink from center outward to pure white on margin.

Ferdinand Mangold. This is the grandest dark Carnation ever seen; flowers large and perfectly formed; color a brilliant red, shaded maroon.

Pride of Kennet. Flowers rich crimson, large and heavy, a good bloomer.

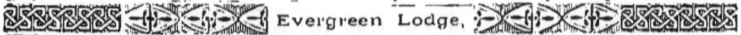

Grace Farden. Flowers medium or large size and very freely produced; plant of grand habit and vigorous growth: superb variety.

Hinze's White. Good, strong, dwarf.

Hinze's Red. Vigorous; very fine.

J. J. Harrison. Flowers a pure satiny white, marked and shaded with rosy pink; good size, of perfect form, never bursts, and is very freely produced on long stems.

John McCulloch. This splendid new Carnation is the most brilliant and finest scarlet yet introduced.

La Purity. Deep carmine.

Lady Emma. Intense scarlet; a profuse bloomer.

La Excellent. White, with carmine edge.

Lydia. Orange and rose; a very free and profuse bloomer.

L. L. Lamborne. Flowers large of purest white, dwarf habit.

Mrs. McKenzie. A very large pink.

Mrs. Carnegie. Color pure satiny white, with beautiful rosy pink stripes; flowers extra large.

Petunia. This resembles a double Petunia as to be most appropriately named; the flowers are large, of a lavender rose, mottled white and deeply fringed.

Portia. The most intense bright scarlet; the flowers are of small size, but of fine shape and long stemmed, and freely produced.

Peter Henderson. Pure white, very fine.

President Degraw. Very fine white.

Quaker City. Magnificent hardy white very profitable for Spring forcing.

Sunrise. Orange, flaked with crimson, a new variety.

Snow White. Pure white.

Snowden. Free blooming; white.

Silver Spray. White, flowers large, long stemmed and freely produced.

MOONFLOWERS.

WE will send two plants, one of each Ipomœa Pandurata and Ipomœa Noctiflora, the Evening Bloomer and the Day Bloomer, for 35 cents.

Ipomœa Pandurata. The Hardy Moonflower. We have annually a great demand for a vine that is hardy that will grow quick, and flower abundantly, the first year it is planted, and we take pleasure in offering Ipomœa Pandurata to meet all these requirements. It is hardy as far North as Boston, will grow 25 feet high the first year it is planted, and being a hardy perennial will come up each year and produce beautiful flowers and a dense shade each season without any further care or expense. Grown with Ipomœa Noctiflora this day bloomer is bound to create a sensation, enabling anyone to have Moonflowers (?) day and night. The color is white, shading to pink and purple in the throat, and its blooming period lasts through several months. The stock we have grown are all nice tubers, varying in shape like sweet potatoes, and will begin to grow at once when planted. Price of strong tubers, 25 cents each.

Ipomœa Noctiflora. The Evening Glory, or Moonflower. There are few plants we have seen sent out that have been so satisfactory as this: hundreds to whom we have sent it have written to us about the satisfaction it has given. One lady says that it was trained on strings to a balcony twenty-five feet high and forty feet wide, and that from August to November it was covered nightly with its white moon-like flowers from five to six inches in diameter. It has also a rich, Jessamine-like odor at night. 15 cents each.

VERBENAS.

WE have a large stock of these useful and popular bedding plants, and grow them extensively. The following comprise the best and most distinct colors of the new Mammoth strain, the

distinguishing peculiarity of which is that the flowers are very much larger than the ordinary type, each individual floret being of the size of a silver quarter dollar, and the truss fully nine inches in circumference; they are of all the shades known to Verbenas. Price 10 cents each, sixteen for $1.00, purchaser's selection. Our selection, by mail, eighteen for $1.00; by express, twenty for $1.00.

Auricula. Fine large purple.
Admiration. A rich clear vermillion; large white eye; extra.
Blue Bonnet. Rich deep blue.
Beauty of Oxford. Dark pink, immense size.
Bernica. Crimson maroon; good flower.
Bijou. Rich dazzling scarlet, small white eye.
Blue Bird. Blush purple.
Candidissima. Finest white.
Columbia. White, striped purple.
Century. Rich dazzling scarlet.
Coral. Fine coral pink.
Crystal. Pure white.
Damson. Rich purple mauve, with clear white centre.
Daisy Dale. Beautiful pink.
Endymion. Deep vermilion, crimson shaded, large white eye; extra.
Fanny. Violet rose, large white eye.
Flame. Bright dazzling red.
Glow Worm. Brilliant scarlet, perfect form.

Maltese. Lilac shaded blue.
Mrs. Massey. Salmon pink, large white centre.
Marion. Mauve, of perfect form, white centre.
Miss Woodruff. Dazzling scarlet, very fine.
May Queen. Soft magenta pink.
Niobe. Deep vermilion, fine flower.
Nelly Park. Orange scarlet, splendid.
Purple Queen. Royal purple, with large white eye.
Perfection. A rich chocolate maroon, lemon eye.
Rosy Morn. Pink, with large white eye.
Striata. White and purple streaks.
Sylphe. The best white Verbena in cultivation.
Snow Flake. Pure white; large truss; a fine and healthy grower.
Scarlet King. A fine, vivid scarlet; dark eye.
Surprise. Clear, orange scarlet, white eye.
Undine. Clear cinnabar-red; extra.

LILIES.

LILIES have long been celebrated for their chaste and rare beauty. It always has been and always will be a favorite. Its name has been handed down to us from the most remote ages, immortalized by painters and poets as emblematical of purity and beauty. No plants capable of being cultivated out of doors possesses so many charms; rich and varied in color, stately and handsome in habit, profuse in variety, and of delicious fragrance, they stand prominently out from all other hardy plants, and no herbaceous border, however select, should be without a few of its best sorts. During the months of February and March, we can send by express Lilium Harrissii, grown in pots, with stems from one to two feet high, fine healthy plants, for 50 cents each, that can be had in bloom at any desired time, according to the size of the plants selected.

Lilium Auratum. Gold banded; the finest of all. 25 cents each; $2.50 per dozen.
Lilium Candidum. The white Lily. 15 cents each; $1.50 per dozen.
Lilium Davaricum. Red tinged, yellow spotted. 25 cents each; $2.50 per dozen.
Lilium Davaricum Incomparabills. Is spotted with crimson. 25 cents each; $2.50 per dozen.
Lilium Longiflorum. Pure white. 25 cents each; $2.50 per dozen.

Lilium Lancifolium Roseum. Blotched white and rose. 25 cents each; $2.50 per dozen.
Lilium Martagon. Purple. 25 cents each; $2.50 per dozen.
Lilium Harrissil. The Bermuda Easter Lily. 25 cents each; $2.50 per dozen.
Calla Lily. Strong plants. 25 to 50 cents each.
Lily of the Valley. Strong clumps. 25 cents each.

HOW TO MAKE A LILY POND.

The best way to grow Lilies and other Aquatics for ornament, is to dig a place two feet deep and as large as you wish it, cement it, and divide it into compartments four by four (to keep them from spreading) by building walls one foot high across it with brick. Place good rich soil in this, and set one plant in each compartment. Set Bananas around the sides, and it will prove the greatest attraction you can grow, and well worth the labor bestowed upon it.

WATER LILIES IN TUBS OR CEMENT BASINS.

For the open air a good degree of success may be attained by planting them in large tubs or half-barrels, on the surface or sunk in the ground. They should be placed where they will receive the full benefit of the sun for at least the greater portion of the day. Fill about half full with the soil recommended for Water Lilies. The next best arrangement for growing the Lily is to build of brick and cement a basin two feet deep and as long as you may desire, either round or square, with a convenient means for emptying the tank at the bottom.

Sagittaria Variabilis. The Arrow Head. A native plant suitable for shallow water, growing about two feet high, bearing arrow shaped leaves and pearly white flowers. 25 cents each.
Nelumbium Leteum. There is scarcely any difference between this and Nelumbium Speciosum except in the color of the flowers, which is of a rich sulphur yellow. A large patch of them, with hundreds of flowers and buds, is a sight never to be forgotten. Tubers, $1.00.
Nymphæa Oderata. Its lovely pure white flowers are worthy of a place beside the most costly Exotics. It can be successfully grown in a tub and Wintered in a cellar. Does well in one of the beds in the Lily tank, but a more satisfactory way than either is to naturalize it in a pond. Strong roots 40 cents each, three for $1.00.

Nymphæa Scutifolia. The Lilies cultivated under these names are of a beautiful shade of lavender blue, not a deep blue, about three or four inches across, but when the plant is given abundance of room and rich soil the flowers will be much larger, and of a decidedly deeper tint. They are very fragrant, the perfume being entirely distinct from that of Nymphæa Oderata. $1.00 each.
Nelumbium Specissum. It is the Sacred Lotus of India and China, and is also cultivated in Japan. 25 cents each.
Nymphæa Oderata Rosea. This is the famous Red Water Lily of Boston; produces flowers larger than the white as commonly seen. It is very fragrant, and in every respect like Nymphæa Oderata except in color, which is a deep pink shade, like Hermosa Rose. $2.00 each.

Pontederia Crassipes. A very interesting tropical aquatic. The leaf-stalk is very much swollen, or inflated, enabling the plant to float on the surface of the water. The flowers are large, of a purplish-lilac shade, and borne on spikes six to eight inches high. It flowers very freely, and is a rapid, spreading grower. It is very ornamental grown in a tub of water on the lawn. It is a native of Guinea, and not hardy, but it can be easily kept in the house in Winter, in a jar of water, or grown in a pot, the same as a Geranium, if the soil is kept very wet. 50 cents each.

Pontederia Cordata. Another interesting plant for shallow water, with heart shaped leaves and spikes of blue flowers, produced all Summer. 25 cents each; $2.00 per dozen.

CLEMATIS.

OF all the hardy running vines in cultivation, none is more beautiful than the Clematis, being entirely hardy and growing as they do more beautiful each year after being planted. They should be grown extensively. To anybody that has a position where a vine can grow, by all means, we say, plant a Clematis, for they are truly not only "things of beauty, but a joy forever." Large strong plants, 75 cents each.

Azure. Light blue.
Aurora. Double red, shaded mauve.
Albertina. Double flowering white.
Duchess of Edinburg. Double white.
Duke of Teck. White and mauve.
Gipsy Queen. Rich velvety purple.
Fair Rosamond. Blush white, red bar.
Gem. Deep lavender blue.
Hybrida Splendida. Reddish violet.
Helena. Pure white, colored anthers.
Jackmanni. Intense violet purple.
Jackmanni Superba. Extra large very dark purple.
Jeanne de Arc. Grayish white.
John Gould Veitch. Rosette: lavender.
Lady Caroline Neville. French white.
Lady Londesborough. Gray, with pale bar.
Lanuginosa. Pale lavender.
Lord Londesborough. Mauve, red bar.
Lucie Lemoine. White, yellow anthers.
Madame Granger. Purplish red.
Madame Torreana. Bright rose.
Otto Frœbel. Grayish white.
Rubella. Rich scarlet purple.
Rubra Violacea. Maroon purple.
Standishi. Light mauve.
Sophia Flora Plena. A double flowering, lilac purple.
The Queen. Fine mauve, Linuginosa-like.

GERANIUMS.

THERE is hardly a plant which is more popular among all classes on the globe than what is generally known as the Horse-Shoe, Zonale, or Fish Geranium. The Geranium is found under many different circumstances; it helps to embellish the conservatories of millionaires as well as the homes of the humble and industrious, but it loses nothing of its inherent beauty on that account.

It wanders with the household furniture from place to place, and the good wife makes a special request to her husband not to forget her Geranium. The propagation of the Geranium is universally known. Every woman knows how to slip it or grow it from cuttings; it is also produced freely from seed. They stand the hot sun of the South better than any other class of plants. They produce more flowers and make a better display on whatever place they are grown than anything that could be grown on a similar space. Our stock is very fine, and much larger than we ever sent out before.

An Extra Choice Assortment.

The following varieties are new and very beautiful. They attract attention wherever seen, by their rich and distinct colors. Price 10 Cents Each, Fourteen for $1.00, by Mail. Our Selection, Sixteen for $1.00, by Express.

Euginirer Cleveland. Single crimson, shaded purple; large.
Julius Lartique. Single rose, vigorous and free.
Tunisse. A double salmon, very handsome.
Countess Ettinger. Single violet carmine.
Francois Argo. Single peach and salmon tinted.
Leon Perrault. Finest single scarlet.
M. Jasaine. Double pink.
La Cid. Fine double crimson.
M. Press. Beautiful double salmon.
M. Carr. Double dark crimson.

Mrs. L. Durant. Single crimson, very free and good.
Mat Sandorph. Semi-double salmon.
Queen of Belgians. Single white.
Gardner Gardett. Single purplish crimson.
Louis Ulbach. Large single scarlet.
Geo. Pardius. Double violet crimson.
Alphonse Daudet. Fine rosy salmon; very attractive.
Annie Gaubon. Double rose.
Kate Schultz. Finest salmon pink, beautiful.
Walter Scott. Rich dark crimson.

Double Flowering Varieties.

Price 10 Cents Each. Our Selection of Varieties, by Express, Twenty for $1.00; by Mail, Sixteen for $1.00. We have a number of fine well established plants in three and four-inch pots at 20 cents each; $1.75 per dozen.

Aimee Goubin. Violet crimson, scarlet shading, very large individual florets, of superb form.
Asa Gray. A light salmon dwarf; very free flowering.
Amelia Baltet. Best double pure white.
Bac-ninh. Immense trusses of large florets, centre beautiful salmon, petals bordered with lively red.
Belle Nencienne. Plant dwarf and floriferous, with trusses of large, full florets of a fresh and very attractive color.

Bruant. A grand sort, trusses and pipe of immense size, semi-double, the color a beautiful, brilliant and sparkling vermilion.
Bridal Bouquet. Beautiful double white flowers, producing freely.
Blanc Parfait. Large round petals, perfect formed florets, plant dwarf and free blooming; a most beautiful variety.
Banquise. Very large trusses of beautifully formed flowers, on strong foot stalks; color pure white.

Centaur. Carries the largest and most perfect truss of any of the pink doubles.

Caudidissima. A large, full and finely formed flower of the most snowy whiteness, not changing to pink.

Charles Darwin. Rich deep violet softening; the base of the upper petal marked with red; a vigorous grower and free bloomer.

Cesare Gandola. Flowers very full of peculiar yellowish-red color.

Consellor Galy. Large trusses of a clear, brilliant current-red color.

De Brazza. Plant of free growing and blooming habit; very large trusses of large semi-double florets; color a beautiful madder orange.

Deputy Lafitz. A rich vermillion purple; extra.

Deputy Varnay. Rich pink base of petals; white.

De Torry. Beautiful shell pink.

Earnest Lauth. Color deep violet; extra large truss.

Ethel Beal. A rich pink shaded carmine.

Etoile des Roses. Color bright, beautiful china rose, base of petals pure white, truss extra large and finely formed.

Gloric de France. Large round flowers; color salmon white.

Gertrude. Color bright salmon, with centre and outer edges touched with white.

Grand Chancellor Faidherbe. A new sort, very thick and double flowers, of a dark soft red, tinted with scarlet and heavily shaded with maroon.

Gilded Gold. Bright orange scarlet of flame color; flowers large, of fine form, branching habit; a very constant bloomer, and one of the best.

General Millot. Large florets, trusses full, of immense size; color the same as that splendid old variety, Grand Chancellor Feidherbe.

Gustave Wideman. A plant of short jointed and free blooming habit; strong trusses of very large florets of a lively apricot color.

Hon. W. Bealby. Red garnet reflecting to violet color; plant of good habit.

Le Sid. Very compact, but vigorous in growth and of perfect habit; color brilliant crimson red.

La Prophete. One of the most magnificent double scarlets ever grown.

L'Anne Terrible. A blazing scarlet.

La Constitution. A glowing yellowish salmon, the nearest approach to yellow of any.

Le Negro. A rich maroon, the darkest variety we have seen.

La Fraicheur. Plant short jointed and of very free growth, freely producing very large trusses of well formed flowers of a tender lilac rose; a new shade of color and quite distinct.

L'Eprouve. Semi-double flowers of a clear carmine; changing to dark carmine, base of petals a pure white; plant short jointed and free blooming.

Lolita Pena. Long peduncles; very large and semi-double flowers, of a lively magenta color.

La Victoire. Trusses large, flowers full and of fine form; color pure and constant white.

La Traviata. Large florets and trusses which are freely produced, carmine violet, upper petals marked with fiery red.

La Vienne. Dwarf and short jointed, creamy white, semi-double.

L'Andalouse. Large trusses, pure white and beautiful flowers, which are freely produced; a plant of very free growth and of fine habit.

Marvel. Dark crimson maroon, an extra fine variety.

Mrs. E. G. Hill. A ground color pale blush, overlaid with a delicate lavender shade.

Medora. A beautiful scarlet, shaded amaranth.

Mrs. Charles Pease. A beautiful, distinct variety of a deep pink color, the upper petals marked white; a great acquisition.

Mon. Gelein Lowagie. The brightest orange shaded.

Mad. Thibaud. A beautiful rich rose shaded with carmine violet.

Mad. P. Owerin. A beautiful rich magenta; semi-double.

Meteor. Pink, shaded crimson; of a rosette form.

Mrs. W. P. Simmons. The flowers and trusses large, deep salmon with deep bronze shadings; plant of free blooming habit.

Marquise de Oysonville. Very compact trusses of the richest carmine Chinese varnish color.

Mon. Jules Aldebert. Flowers semi-double, of a beautiful flaming capucine orange color.

M. Roche Alex. Large flowers, the lower petals a rosy salmon, centre and upper petals fiery salmon; distinct and beautiful.

Mad. Ed. Andre. Large umbels, full and well formed florets, color salmon with bronze shadings.

M. Hardy. Large flowers, a deep lilac and tender rose.

Mad. Hoste. Perfectly formed trusses of enormous size; florets very large and nicely formed; color a novel shade of tender salmon, bordered with white.

M. Henri Truchet. We have an improved L'Eprouve in this variety; color clear carmine, base of petals pure white; a splendid variety.

Mon. J. Cretien. Enormous trusses of a fiery red velvet color.

Mad. L. de Beuregard. Enormous trusses of very large double florets of the finest form; color a lively salmon, each petal distinctly bordered with white.

Naomi. Blush pink shaded white, lovely color; one of the best.

Orange Perfection. Enormous trusses of orange vermillion colored flowers.

Palmyra. Immense sized trusses of well formed florets; color white; a fine bedder.

Panama. Enormous trusses borne on strong foot stalks; flowers full and well formed; centre of petals salmon vermillion, bordered with rosy salmon.

Prokup Daubek. Very bright, soft rose color, a most charming shade; a very beautiful flower.

Peter Henderson. A very fine variety with bright orange scarlet flowers of fine shape.

49

President Leon Simon. Flowers large and of perfect form; bright clear red shaded salmon; truss very fine and large.

Peach Blossom. Semi-double; beautiful rosy peach color; a good bloomer.

Phyllas. Salmon centre; the edge of petals lighter.

Robert George. Deep crimson scarlet of great size; a free bloomer.

Richard Brett. A peculiar orange color, the nearest approach to yellow good bedder.

Solferino. Red, with fine solferino shade.

S. S. Nutt. Rich crimson, dark trusses, massive and profuse.

Telephone. Double scarlet.

The Ghost. A waxy white; good form and substance; very desirable.

Thibaut et Keteleer. Flowers semi-double, salmon, bordered vermillion.

Venus. A creamy white; large truss.

Wagner. A fiery carmine; large flower and truss; a profuse bloomer.

Single Flowered.

Arc-en-Ciel. The trusses of this variety exceed in size nearly all the single-flowered; color lake red, upper petals marked with orange scarlet.

Arizema. Pure white, shaded with a delicate pink.

Bridesmaid. A light pink, dotted with brighter pink.

Beauty of Clarksville. A large rosy scarlet, with white eye; one of the best bedding varieties.

Cyclope. Trusses large; color white, shaded salmon, with an orange centre plainly distinguished from the white.

Cosmos. Immense perfect formed trusses; florets large and finely formed; color salmon, with orange, brighter towards the centre.

Constellation. A dazzling scarlet with white eye; a splendid bedding variety.

Colonel Holden. A very beautiful rosy crimson; a distinct color, free bloomer.

Dragon. A ceris scarlet; very large truss and flower.

Eva. White, with salmon centre.

Electrician. A beautiful shade of rosy salmon; a delicate and novel shade.

Fairy Queen. Salmon centre; outer edge of petals white.

Favourite. Florets large and fine, truss of immense size, color beautiful sherry scarlet; plant of good habit and free blooming.

General Grant. Scarlet; a good bedder.

George W. Earl. Pure white, with broad, deep pink centre; beautiful large flowers; a free bloomer.

Girardin. Produces immense trusses; a clear color.

General Sheridan. Brilliant crimson; of dwarf habit; large truss.

Heleranthe. Light gold dust red; cup shaped flowers; extra.

Harry King. A magnificent Zonale; a a vivid crimson scarlet; extra.

Hilda. Pale rose, white centre; a delicate shade.

Jules Ferry. Splendid trusses on long rigid footstalks, which are held well above the foliage; scarlet red.

Jumbo. Florets and trusses of immense size, of a rich deep crimson color.

Jean Sisley. The richest scarlet.

James Vick. Flowers and trusses of a great color, a deep flesh, with dark bronze shadings; of free habit.

Louis Uhlbach. Intense dazzling scarlet.

Lady Byron. One of the finest pink flowers with white eye.

Miss Blanche. Deep purplish pink, trusses of immense size and freely produced; this is a grand bedding sort.

Mrs. E. T Keim. Pearly white with vermillion centre; large white eye.

Mary H. Foote. One of the most beautisalmon colored, even shade; very fine.

Mrs. Hamilton. Rosy pink with a fine violet shade.

Minerva. A bright, rosy salmon; very large truss and flowers; a most beautiful shade; very free flowering.

Marshal Valiant. Immense truss; extra large flowers; color rosy salmon; distinct.

Mrs. James Vick. White edges; pinkish centre, it is without an equal for for Winter.

Mrs. Moore. Pure white, with a beautiful ring of bright salmon around a small white eye; of a dwarf habit; free flowering; very desirable.

Master Christine. Pink; a good bedder.

New Life. A sport from striped Vesuvins, having its bright scarlet flowers striped and flaked with salmon and white.

Neve. Plant vigorous and of splendid habit; large trusses of the purest white.

Poet National. Round florets nicely displayed; color of Baroness Rothschild rose, deeping to soft rosy peach.

Queen of the West. Bright orange scarlet; very large truss; profuse bloomer; we know of no finer for planting out in beds.

Ralph. A fine dark crimson, suffused with amaranth; large, well formed truss; very distinct; extra.

Rev. W. Atkinson. A fine dark crimson scarlet; very large flower and truss.

Reflector. Very bright and handsome scarlet, with large pure white eye; trusses large and freely produced.

Rosy Morn. Beautiful well formed and bright rosy carmine flowers; of a neat, compact habit.

Rosemond. A vermillion scarlet, beautifully shaded with rose.

Swanley Gem. An English variety of exceptional merit; color rosy salmon red, with large white eye.

Sam Sloan. A fine bedding variety; color deep velvety crimson, large truss and very free flowering.

Starlight. Pure white with broad pink centre and distinct pure white eye.

Senator. A dark scarlet; very large truss.

Triumph. Rosy salmon, shaded pink; very large flower and truss; distinct and floriferous.

Victor Hugo. Plant dwarf and floriferous, trusses large, flowers fine form and finish, a brilliant salmon.

50

Sheen Rival. A rosy scarlet; beautiful stems; zone creamy white.
Voltaire. A large crimson scarlet; truss very effective.
Wood Nymph. Light pink; very free blooming variety.

Wm. Cullen Bryant. The finest shaped single flowering Geranium known; color a soft, rich, pure scarlet.
White Swan. A large bold white flower fine for bedding.

Ivy Leaved.

Bijou. Hybrid; double scarlet.
Dolly Varden. Gold and bronze.
German or Parlor Ivy.

Remarkable. Flowers rose and white; strong growing variety, suitable for hanging baskets. 15 cents each.

Gold, Bronze and Silver Leaved.

Bijou. Flowers a dazzling scarlet; leaves bordering white.
Black Douglass. Yellow, with dark zone.
Battersou Park Gem. Yellow and green.
Golden Harry Tricolor. Golden yellow, bronze zone.
Marshal McMahou. Yellow ground, with a bronze zone.
Mountain of Snow. Flowers bright scarlet, the leaves margined white.

Happy Thought. Entirely distinct from any other variegated Geranium; centre of the leaf creamy yellow, with a broad margin of deep green.
Mad. Sllaeroi. New Silver Geranium this is the greatest acquisition in variegated Geraniums, for bedding purposes, that has been introduced since the old Mountain of Snow.

Scented.

Apple. 25 cents each.
Lemou. 10 cents each.
Nutmeg. 10 cents each.
Oak Leaf. 10 cents each.
Pennyroyal. 10 cents each.

Rose. Large varieties. 10 cents each.
Mrs. Taylor. A distinct new variety of the rose scented Geranium, having large scarlet flowers. 15 cents each.

Pelargoniums.

A fine assortment, 30 cents each.

⁂

Select Hardy Evergreen Trees.

JUNIPERS.
Irish. Erect and formal in its habit; is much used in cemeteries. Eight inches high, 25 cents each.
Hemlock. Remarkably graceful, beautiful tree, with drooping branches and the delicate dark foliage of the Yew. 50 cents each.
Arbor Vitæ, Golden. Beautiful Chinese variety, compact and globular in form, a lively yellowish green. 50 cents.
Arbor Vitæ, Semper Aurea. Of dwarf habit, but free growth, retaining its golden tint the year round. 50 to 75 cents.
Arbor Vitæ, American. This plant is, all things considered, the finest Evergreen. 25 to 50 cents.
Arbor Vitæ, Tom Thumb. Very small, compact little Evergreen; a beautiful ornament for a small yard or cemetery lot. 50 to 75 cents.
Arbor Vitæ, Hoveyi. Small tree, globular in form; foliage light green, with a golden tinge. 50 to 75 cents.
MAHONIA AQUIFOLIA.
Evergreens with bright shiny leaves and showy bunches of yellow flowers in the early Spring. 25 cents each.
MAGNOLIA.
Grandiflora. The true Southern Magnolia; of great beauty; too wellknown to need description. Nice pot, plants, sure to grow, about 18 inches high. 75 cents. This size is much safer to plant than the larger sizes.
BOX.
Dwarf. Fine for edging. 10 cents. $1.00 per 100.
SPRUCE.
Norway. Lofty elegant tree of perfect pyramidal habit, very popular, should be largely planted. One of the best Evergreens. 50 cents.
Arbor Vitæ, Pyramidalis. Exceedingly beautiful, bright variety, resembling the Irish Juniper in form. 50 to 75 cents.
Retinospora Plumosa. Exceedingly handsome Evergreens from Japan, with feathery, light green foliage. 25 cents.
Retinospora, Plumosa Aurea. Like the preceding, a plant of great beauty, soft, plume-like foliage. 25 cents.
Yew, Irish. Upright in growth, with dense foliage of a dark sombre hue; valuable for cemeteries or small yards. One dollar; strong young plants, 50 cents.
SIBERIAN ARBOR VITÆ.
The best Arborvitæ for this country. Exceedingly hardy, keeping its color well in Winter. 25 and 50 cents each.

DAHLIAS.

Price for Strong Tubers 15 Cents Each, $1.50 Per Dozen.

Young green plants, ready April 1st, 10 cents each, $1.00 per dozen. The tuber or roots cannot be sent by mail.

Show Flowers.

Alderman. Light shaded purple.
Annu Neville. Pure white; extra.
Alexander Crammond. Shaded maroon.
Amazon. Yellow, with scarlet edge.
Ada Tiffin. Light peach, splendid form.
Aristides. Deep purple.
Burgundy. Rich shaded puce, very large.
British Triumph. Rich crimson.
Bob Ridily. Bright red, splendid form.
Burning Coal. Bright scarlet, fine.
Conflagration. A bright orange, tinted scarlet.
Coronet. Blush white, lilac tinted.
Cremorne. Yellow, tipped red.
Cochineal. Crimson, toned with a brownish shade.
Constance. Flowers are of purest white and good shape.
Duke of Wellington. Purple, very large.
Duke of Roxburgh. Salmon buff, extra.
Earl of Shaftsbury. Rich purple, fine.
Earl of Radnor. Deep crimson.
Earl of Beaconsfield. Rich plum, finest form and outline.
Emily. Light purple, very fine.
Edward Purchase. A beautiful bright ▼crimson.
Estella. Cream white.
Fire King. Fiery crimson scarlet.
Flamingo. Deep vermillion scarlet.
Firefly. Orange scarlet, fine.
Glow Worm. A handsome yellow, tipped with cinnabar.
Glare of the Garden. Deep red, large and showy yellow centre, free and effective.
Geo. Smith. Bright magenta, fine form.
George Goodall. Scarlet, most desirable.
Her Majesty. White, deeply edged with purple.
Henry Glasscock. Buff striped, spotted crimson.
Henry Walton. Yellow ground, heavily ▼edged vermillion.
Hercules. Yellow, striped red, frequently ▼self of a brownish red.
Henry Bond. Rosy lilac, large and fine.
Jennie Grieve. White ground, edged a rosy lilac.
John Wyatt. Crimson scarlet, fine form.
James R. Service. Deep yellow.
Jourezi. Dazzling scarlet, long petals, curiously twisted at the points.
John McPherson. Rich violet purple.
John Sladden. Nearly black, extra.
John Harrison. Dark crimson.
John Kirby. Yellow buff, compact.

James Wilder. Rich velvet maroon, red shading.
James Crocker. Fine purple.
King of Primroses. Primrose yellow.
King of the Dwarfs. Deep violet purple.
Little Bride. White, very small and free.
Little Beauty. Pure white.
Lady Mary Herbert. Yellow, tinted buff.
Lady Jane Ellis. Creamy white, tipped purple rose.
Lady Mary Wilde. White, tipped rosy purple.
Livonia. Lilac.
Mary Keyros. Fawn, ground edged a bright rosy purple.
Mrs. Burgess. Purplish shading, occasionally white tipped.
Mrs. Brunton. Pure white ground, laced with deep purple.
Mrs. Goodwin. Large dark maroon.
Mrs. Standcomb. Canary yellow, very distinctly tipped with deep fawn.
Mrs. Piggott. Pure white, of good form, fine.
Masterpiece. Rosy purple, large, finest form.
Mrs. Fordham. French white, tipped with soft purple.
Mrs. Gladstone. Pink.
Modesty. A beautiful shade of yellow, tinted pink.
Mrs. Hawkins. A rich sulphur yellow, light towards the tips.
Netty Buckell. Light blush, tinted pink.
Oreole. Golden yellow, first-class.
Our Tim. Buff, shading to peach.
Oracle. Deep yellow, heavily striped a bright crimson.
Princess. White, large, full and fine.
Princess Alice. Light lilac, extra fine.
Paradise. William's. Clear claret; new in color.
Paul of Paisley. The finest lilac.
Pearl. Pure white; dwarf.
Purity. Purest white, free and constant.
Rifleman. Crimson scarlet, constant.
Royally. Golden yellow, crimson tipped.
Rosette. Rose, edged lilac.
Sir Joseph Paxton. Yellow, tipped red.
Tom Green. Maroon, tipped white.
The Pet. Dark maroon, tipped white.
Unique. Pure white.
Vesta. Purest white, very fine.
Wm. Kynes. Orange, one of the finest formed flowers.
W. Pringle Laird. The finest maroon large, beautifully formed.

Single Dahlias.

Single blooming Dahlias have become very fashionable, owing to the value attached to the cut blooms, their light, butterfly-like forms giving the flower a special grace. The brilliant red is the most showy. 15 cents each.

52

BEGONIAS.

THIS class of plants is each year becoming more deservedly popular. The beauty of their foliage and graceful flowers make them useful plants for greenhouse or window decoration. A number of new varieties of special merit are coming out each year. We would call special attention to Semperflorens Gigantea Rosea as a beautiful decorative plant; also Sutton's White Perfection, a handsome free-flowering variety. Our stock of Begonia Rex is large this year, and contains some handsome and beautifully marked varieties. We will send by express twelve fine well grown plants of Begonias for $1.00, to consist of one Rex, one Semperflorens Gigantea Rosea, one Sutton's White Perfection, and one Metallica, the rest to consist of the other varieties named below, the selection to be left to us. This will make a fine assortment of this popular plant to begin with, and will repay the outlay. We can send twelve smaller plants as above by mail for $1.00.

Alba Picta. A perfectly distinct variety. Leaves are glossy green, thickly spotted with silver white, the spots graduating in size from the centre toward the margin; flowers white. 15 cents.

Begonia Rex. The Rex varieties, of which we have a dozen or more, varying in color and markings, are very effective as pot plants. 25 cents.

Ctssc. Eouise Erclody. Rex variety. This is the Begonia of all Begonias. Its striking peculiarity, which distinguishes it from all Begonias, consists in the two lobes not growing side by side, but one winds itself in a spiral way repeatedly over itself. 35 cents.

Dreggi. This variety is always in flower. Winter and Summer; it is one of the most useful plants we have; flowers are white. 15 cents.

Fuchsioides Alba. Fuchsia-like, pure white flowers. 10 cents.

Fuchsioides Rubra. Red flowers; very fine and constant bloomer.

Incarnata Metallica. Dark green leaves with silver dots and metallic shade, fine pink flower clusters. 10 cents.

Nitida Alba. A strong grower and profuse blooming variety, producing immense panicles of pure white flowers; fragrant. 10 cents.

Parvifolia. A dwarf, bushy growing variety, with pure white flowers, being in bloom the whole year. 10 cents.

Rubra. One of the finest Begonias in cultivation; its dark, glossy, green leaves, combined with its free flowering habit, make it one of the very best plants for house or conservatory decoration; the flowers are of a scarlet rose color, and are produced in profusion. 15 cents.

Hybrida Multiflora. Flowers rose, it blooms almost continually. 10 cents.

Semperflorens Gigantea Rosea. A superb variety, strong, upright growth, fine large flowers of a clear cardinal red, the bud only exceeded in beauty by the open flower, which is borne on a strong, thick stem; the leaves are smooth and glossy, and attached to the main stem; both leaf and stem quite upright growing, forming a shrubby, round plant. 25 cts.

Saundersouil. Flowers a scarlet shade of crimson, borne in profusion during the entire Winter months. 10 cents.

Weltoulensis. Handsome Winter flowering variety, lovely pink flowers, easy cultivation. 10 cents.

Zebrina. Erect, of beautiful variegated foliage, leaves shaped like that of Rubra, and bearing white flowers. 10 cents.

Glaucophylla Scaudens. An early flowering and vigorous growing variety, producing its beautiful clusters of salmon colored flowers from the axil of each leaf; its drooping habit makes it a very desirable plant for hanging baskets. 15 cents.

Metallica. A shrubby variety, a good grower and free bloomer; leaves triangular, longer than they are wide, under side of leaves and stem hairy, the surface of lustrous metallic or bronze color, veined darker; flowers white, covered with glandular red hairs; it is perfectly distinct. 15 cents.

Ingramii. One of the best Winter flowering varieties; flowers reddish carmine, leaves edged with bronze. 10 cents.

Richuifolia. Has large palmated leaves, supported on stems from three to four feet long. 10 cents.

Sutton's White Perfection. A beautiful dwarf, free flowering plant, always in bloom, and attracts attention wherever seen. 25 cents.

TUBEROUS BEGONIAS.
The Flower of the Future.

Recognizing the rapid improvement, magnificent coloring, large size, and effectiveness of this most wonderful flower which is just beginning to be appreciated, on account of its easy culture, as the best of all Summer bedders, we have decided to take up its propagation as a specialty for the reason that during the Summer months it proves to be the most satisfactory of bloomers, vieing with and surpassing the Geranium in brilliancy, and remaining in full flower until cut down by Autumn frosts. Probably no family of plants has been hybridised with such success as this, the result being flowers, both double and single, of innumerable shades of color, ranging from pure white through rose and pink to intense crimson and fiery scarlet, and from the deepest yellow to tawny brown and brilliant orange.

SINGLE.

50 Cents each: $5.00 per Dozen.

A. Mayes. Rich crimson, medium size of great substance, free.

A. W. Tait. Fine flowers, intense crimson, dwarf and free.

Crepuscule. Light buff, edged crimson.

Countess of Bessborough. Clear rich golden yellow, distinctly edged soft red, splendid habit, distinct.

Earl of Bessborough. Clear yellow buff, edged soft red.

Excelsior. Clear madder, bright yellow centre, well formed flowers.

Glow Worm. Intense fiery crimson.

Gigautea. Crimson, petals nearly four inches broad, almost circular and of great substance.

Henry Cannel. Rich magenta rose.

Iudian Chief. Of great substance, of a decided shade of Indian yellow.

J. Downs. A yellow of good habit and substance.

King of the Crimsous. Rich crimson, shaded maroon.

King of the Begonias. Strong, bushy, dwarf habit, intense scarlet-crimson.

La Caudeur. White, fine, dwarf habit, flowers large and well formed.

L'Abbe Froment. Bright golden yellow, erect.

Miss Maleomson. Large white flowers of fine substance.

Mrs. Grove. Good, dwarf, bushy habit; flowers medium size, and of the purest white.

Mrs. Shepherd. Of circular form, erect flowering, pure white.

Magog. Dark scarlet, almost circular, petals of great size.

Mrs. Mauby. Yellow, shaded buff.

Mrs. H. Coppin. Clear yellow, medium sized flowers.

Mr. Cockburu. A fine variety, Orange scarlet, erect flowers, large.

Mrs. H. G. Murry-Stewart. Rich scarlet, robust habit. For exhibition one of the best.

Miss Cannell. Rosy pink, suffused with purple.

Mrs. Bellew. A fine pink.

Mrs. Edwards. Light pink, very large.

Norma. Reddish magenta, large.

Primrose Perfection. Habit and form better than the old varieties.

Queen of Yellows. Pure yellow, deep in shade.

Riley Scott. A rich colored crimson.

Shirley Hibberd. Brilliant scarlet of a dazzling shade, good habit.

T. Moore. Salmon red, good habit.

Total Eclipse. A rich, self-colored.

W. E. Gladstone. Immense fiery crimson, good strong habit.

Wonder. Deep yellow buff, orange yellow at base of petals.

Yellow Bedder. Pure yellow, very free.

We offer a good stock of unnamed seedlings, fine for bedding out or potting, at 25 cents each, $2.50 per dozen, either dormant tubers or growing plants.

DOUBLE.

75 Cents Each.

Antoinette Guerin. Full flowers, broad.

Anna, Countess of Kingston. Salmon, shaded yellow in centre.

Berenice. Bright coral red, rose centre.

Com. Basset. Large, soft red.

Clovis. Orange scarlet full.

Countess of Dudley. A most useful variety, white, tinted cream, free.

Dr. Baillon. Deep salmon.

Eclat. Bright orange scarlet.

Eugene Lequia. Scarlet orange, tipped scarlet.

Etna. Reddish scarlet, strong habit.

Esther. Crimson, rosy pink centre.

Francis Buehner. Clear reddish cerise, shaded orange.

Flugnrant. Deep crimson, free.

Gabrielle Legros. Clear sulphur white, large and imbricated, fine.

Incendie. Suitable for baskets, A rich reddish scarlet.

J. Walker. Crimson scarlet, large.

King of the Yellows. Bright color.

Louis d'Or. Free and effective distinct shade of yellow.

Leon de St. Jean. One of the finest, scarlet red, dwarf.

Louis Bouchet. Brilliant orange scarlet.

Lady Sheffield. Pure white.

M. Rondeau. Soft scarlet, shaded orange.

M. Thiroux. Clear orange scarlet.

M. Duvivier. Rosy crimson, very double.

Major Studdert. Bright red, strong habit.

Mr. G. Hawes. Crimson shaded mahogany, medium size, free.

Mad. Vincenot. French white, centre marked yellow.

Mrs. Webb. Creamy white, good habit.

Major Leudy. Large and fine, distinct shade of pink, mottled soft yellow, with orange centre.

Miss Lucas. Rich salmon, conspicuous guard petals, free and showy.

Mr. Truffaut. Reddish salmon, deeper centre, large fine flowers.

Mad. Grousse. Large nankeen-salmon, distinct.

Mrs. Lewis Castle. Of a beautiful salmon shade, large double flowers, quite distinct.

Mrs. Windsor. Ground color a creamy blush, delicately edged pink, attractive, good habit.

Mrs. Hall. Full size flowers of a creamy shade.

Queen of the Netherlands. White with cream shade in centre.

Robin Adair. Rich glowing crimson.

Sir Bealby. Soft crimson, deeper centre, petals pure white, shaded cream.

Thos. Baines. Deep cream, shaded and edged terra cotta, medium size, free.

Virginalis. Free flowering white.

W. B. Miller. Orange scarlet, good size.

CAMELLIA JAPONICA.

The rich and pleasing contrast afforded by their dark green leaves and their superb flowers of exquisite beauty and waxy texture, together with their almost endless variegations of color, combine to make them one of the most desirable of Winter flowering plants. Price $1.00 each, nice bushy plants about sixteen inches high. We have no large size Camelias to offer this Spring.

Augustina Superba. Of transparent rose color, sometimes spotted with white

Alba Plena. Large flower; whise imbricated.

Archduchess Augusta. Beautiful red, with a dark azure vein and a white band in the middle of each petal, the flower assuming a blush and variegated color.

Angelo Cocchi. White, sometimes spotted or striped bright red, sometimes dark.

Archduchess Marie. Magnificent flower of good form, very double; vivid red, with white rib-bands.

Aspasia. Small petals, very compact, brilliant red, rosy white heart.

Auguste Delfosse. Fiery rose color, centre of petals striped; finely imbricated.

Bella Romana. Flowers good; color soft blush, flaked with crimson.

Bonomiana. Large petals, well rounded, imbricated in regular form; white line crossed through and through with deep red.

Chandleri. Flowers large and petals broad, of a rich pink color.

Comtesse of Orkney. Pure white with carmine stripes, often peony-form in the centre, very large petals, sometimes a very bright red, shaded dark red or white rosy stripes, with the edges pure white.

Commendatore Bettl. Superb variety, finely imbricate; red, chaning to rose.

Comtesse Lavinia Mazzi. Large buds; flower well formed, dotted cherry red.

Candissima. Pure white; imbricated.

Duchess de Berry. Flowers large, pure white; habit good, with fine foliage.

Fanny Bollis. Magnificent flower, well formed, large rounded petals of flesh white, stained blood red.

Imperatrice Maria Theresa. Large and splendid imbricated flower; petals bright red, changing to pure white.

Imbricata. Carmine red, sometimes of variegated color, hence called Imbricata Tricolor.

Jubilee. Extra large flower, imbricated; petals large and rounded; centre white; lightly rose sprinkled.

Leon Leguay. Very double; red, shaded deep red, exterior petals undulated; a first-class variety.

Leopold I. Bright scarlet red, with plush crimson bars near the border of the petals; extra fine variety.

Lyanna Superba. A vivid red, flower imbricated.

Mathotiana Alba. Pure white.

Relue des Fleurs. Small leaves, but a vigorous grower, of good habit; a deep rich crimson color.

Trionfa di Lodi. Imbricated; large white petals, speckled and striped.

Mistress Cope. White flower, crimson stripe; of splendid form; extra.

Mad. Leboise. Imbricated, bright red.

Nobilissima. Pure white; peony-formed; very highly valued on account of its early flowering.

Princess Baciochi. Superb flower, well imbricated; cherry red, with small white bands.

Prince Albert. Blush white, with numerous stripes of deep rose.

Princess Clothine. Imbricated, nearly double; it has strong petals with large white bands and deep red bars.

Reine Marie Henrietta. Of very fine form, of splendid foliage; rose color, often speckled pure white; perfectly imbricated; very free bloomer.

Triomphe de Wondelghem. Flower red, central ribbon more bright, sometimes carmine with rosy white band.

Union. Very large pure white flower, sometimes peony-formed; first-class variety.

AZALEAS.

AZALEAS are a class of plants highly ornamental for Winter and early Spring flowering. They are of easy culture and can be had in bloom from Christmas to May if a fair selection of varieties is kept up. Our Azaleas will be this season, as usual, the very best for shape, variety and bud. Nice shaped plants, 50 cents each; large plants, $1.00 each.

A. Borsig. A very fine pure white and double variety, of good form and great substance.

Alba Illustrata. Flower of purest white, occasionally sprinkled with lilac rose.

Alba Illustrata Plena. Pure white, double, fine for forcing.

Apollon. Pure white, sometimes lined with bright red.

Baronne de Vriere. Flowers enormous, snow white, petals very large, with undulated margin, sparingly striped with crimson, and blotched with sulphur yellow.

Bernard Andrea. Rosy purple, double; very beautiful.

Bernard Andrea Alba. Superb white flower, very double; a most desirable and beautiful variety.

Ceres. White, blotched with rose; a very profuse bloomer.

Charles Enke. Rosy salmon, marginated with white; very fine.

Charmer. Bright amaranth, the upper petals blotched with a deeper shade.

Comte do Chambord. A salmony rose color, striped and edged by a wide festoon of the purest white.

Daphne. Fine large semi-double variety, pure white.

Dr. Moore. Intense rose, with white and violet reflection.

Deutche Perle. Double pure white: very free flowering and early blooming. If placed in a gentle heat it will flower at the beginning of December; it may be regarded as the best among the double-flowered whites.

Etendard de Flandre. White, striped with purple.

Eugene Mazel. Rosy salmon, the upper lobes violet.

Flag of Truce. Pure white, double and very full: one of the finest double white Azaleas in cultivation.

Indica Alba. Single white.

Iveryana. White, with red stripes.

Jean Vervaena. Deep, rich crimson, edged with white, dark spot on upper petals.

Le Flambeau. Glowing crimson, bright and effective.

Kenogin Cleopatra. Beautiful single variety, with exceedingly large flowers: white, spotted and striped with rosy carmine.

Mad. Iris Lefebvre. Flowers extremely double, of a dark orange red, broadly banded and striped with deep brownish violet.

Mad. Paul de Schryver. The flowers are large, well made and very double, having the centre sometimes imbricated like that of a Camellia: lively violet rose.

Mad. Vander Cruyssen. Soft, glossy rose, tinted with amaranth: very large semi-double flower: fine and showy.

Model. Bright rose.

Mad. Dom. Vervaene. Vivid salmon rose, white margin.

Mad'le Leoni Van Houtte. Very fine white, flaked with rose and spotted with sulphur yellow.

Mad'le Marie Lefebre. A large flower of exquisite form and substance: pure white.

Marquis of Lorne. The flowers are of a beautiful orange, with saffron yellow blotch; the petals are very large and round.

Oswald de Kerchove. A very beautiful variety of lake rose with fiery blotch: large and well formed.

Princess Charlotte. Very large flower of a beautiful rose: fine form.

Raphael. Alba Illustrata Plena. Pure white, double: fine for forcing.

Reine des Pays-Bas. Violet pink, margined with white and richly striped with deep crimson.

Rol Leopold. Rich glossy crimson: very fine form

Sigismund Rucker. Flowers lilac rose, strongly netted, bordered with white, splendidly blotched with white crimson.

Superba. Bright rosy carmine, of good form ; late.

Vittata Crispiflora. White, shaded with purple and crimson: very free and fine sor forcing.

W. Wilson Saunders. A very fine white variety, striped and blotched with vivid red.

PALMS.

THE following is a select list of rare and handsome varieties, which can be recommended for apartments, conservatory decoration, or vase plants during the Summer. All are in a clean and thrifty condition suitable for making immediate effects, and require no nursing to bring them into proper shape. The Seaforthias, Areeas, Latanias and Kentias are of quick and graceful growth, and can be grown without much trouble.

Areca Baueri. A distinct and graceful Palm, excellent for table decoration, $2.00 each.

Arecrn Rubra. Foliage deep green, tinged red, stems red. $1.00 each.

Areca Lutescens. One of the most beautiful and valuable Palms in cultivation; bright glossy green foliage and rich golden yellow stems. $2.00 each.

Areca Sapida. A strong upright growing variety with dark green feathered foliage. $1.00 to $8.00 each.

Caryota Urens. An easily grown and useful sort. $1.00 each.

Chamœdora Elegans. A pretty decorative variety with deep glaucous foliage. $1.00 each.

Chamœrops Excelsa. A handsome Fan Palm, of rapid, easy culture. $1.00 each.

Curculigo Recurvata. A very graceful Palm like plant for decorative purposes. $1.00 each.

Cycus Revoluta. The stem of this variety is very thick, and bears the foliage in whorls at the top. $5.00 to $12.00 each.

Kentia Balmoreana. Beautiful strong growing Palm, with deep green crisp foliage. $1.50 to $3.00 each.

Euterpe Edulis. Of spreading graceful habit. $1.00 each.

Kentia Fosteriana. One of the finest of the Kentias, with graceful bright green foliage. $1.50 to $3.00 each.

Latania Borbonica. Chinese Fan Palm. The most desirable for general cultivation, especially adapted for centres of baskets, pases, jardinieres, etc. $1.00, $2.00, $3.00, $5.00 each.

Phoenix Reclinata. Beautiful reclinate foliage. $2.50 each.

Pœnix Rupicola. Of graceful arching habit. $2.50 each.

Phœnix Sylvestris. An attractive sort, deep green foliage. $2.00 each.

Seforthia Elegans. One of the very best for ordinary purposes, of graceful habit, and rapid, easy growth. $1.50 each.

Pandanus Utilis. Screw Pine. Called Screw Pine from the arrangement of the leaves on the stem. Excellent for the centre of vases and baskets, or grown as a single specimen; a beautiful plant. 75 cents to $1.00 each.

Pandanus Veitchii. This is one of the most attractive plants. The leaves are light green, beautifully marked with broad stripes and bands of pure white, and gracefully curved. $1.00.

FUCHSIAS.

THESE, when in full bloom, are the most graceful of all cultivated plants; nothing can surpass the beauty of well grown specimens. They delight in a light, rich soil, and may be grown either as pot plants or in a sheltered border. In either case they should be protected from the hot mid-day sun and from heavy currents of air. They require plenty of water and partial shade. Price 10 cents each, 14 for $1.00, purchaser's selection. Our selection, 16 for $1.00, by express.

Admiral Courbet. Enormous double flowers; corolla deep violet, tube and sepals bright red.

Admiral Miot. Plant very bushy, free bloomer; large double corolla of a clear prune color, sepals brilliant red.

Alice Mary Pearson. Sepals creamy white, tube rather long, single corolla of a dark crimson red color.

Annie Earle. Tube and sepals a waxy white, corolla single, clear carmine.

Carl Halt. Corolla single, bright scarlet, with white stripes.

Clio. Tube brilliant rose, sepals bright red, corolla single, purest white.

Col. Dominie. Of a very free branching habit, corolla very double and of an imbricated form, white striped rose, sepals reflexed and of a distinct, clear red.

Cleopatra. Very large and double corolla of an azure blue color, passing to dark violet; extra fine.

De Mirble. Plant of fine robust habit, sepals a bright red, large single corolla, violet and rose.

Rsmeralda. One of the grandest and most distinct varieties; tube short, sepals red, corolla very large and double, beautiful lilac changing to rose.

Flocon de Neige. Large bell shaped corolla of a creamy white, sepals clear carmine, plant dwarf, and of freest blooming habit.

J. J. Rosseau. Very large and full corolla, a bluish violet, sepals bright red; in fact a very fine plant.

Luster. Tube and sepals white, corolla scarlet.

Mrs. E. G. Hill. Undeniably the most perfect and beautiful double white ever raised; the short tube and sepals are a bright, rich reddish crimson color, the corolla extra large, full and double.

Mazeppa. Very free bloomer, flowers single, sepals relieved, fiery red, corolla violet red.

Mde. Von der Strauss. Flowers large, sepals slender and well refllexed, and of a pure red; corolla white, large, double.

Mon. Thibaut. Vigorous and remarkably free blooming, tube stout, sepals dark red, corolla rose vermillion, tinted violet.

Penelope. Grand single white, semi-double, corolla long, large, of beautiful form, purest white, sepals lively red.

Parmentier. Sepals a coral red, corolla violet, round and double.

Perle Von Bruun. Sepals recurved, very clear in color, of immense size, and has a double corolla of the purest white.

President F. Gunther. Corolla double, lilac and violet.

Phenomenal. Without doubt the largest purple Fuchsia yet produced; corolla a purplish lilac, sepals beautiful coral red.

Purple Prince. Double purple corolla.

Regent. Sepals recurved, a violet carmine, corolla double, violet and blue.

Rose of Castile. Corolla violet, sepals white, of splendid habit.

Snow Fairy. Double white corolla.

Storm King. The king of all the white Fuchsias; although it is claimed by some to have flowers as large as tea cups, we do not make such mistakes, but we say it is of a very large flowering variety, but of a very dwarf growth.

Speciosa. Pale ree tube and sepals, dark red corolla; best Winter bloomer.

Sea Berlet. Carmine tube and sepals, corolla of immense size, double, deep violet purple.

Ville de Lyon. Tube large, sepals horizontal, crimson red, corolla large and double, white veined carmine.

GENERAL COLLECTION OF PLANTS
Suitable for Greenhouse or Out-Door Culture.

AURICARIA EXELSA. A beautiful evergreen tree of handsome uniformity of growth, the branches produced at right angles all the way up the main stem; must be seen to be appreciated; very scarce. Small plants, $4.00 each; nice strong plants, $8.00 each.

AURICARIA IMBRICATA. Known as the "Monkey Puzzler," being covered with sharp thorns. It is said it is the only tree the monkey cannot climb. Small plants, $1.00 each; large plants, $5.00 each.

ANTHERICUM VARIEGATUM. A most striking novelty, introduced from the Cape of Good Hope. Very valuable as a decorative plant, being suitable either for the greenhouse, parlor, or dining table. The foliage is of a bright grassy green, beautifully striped and margined with a creamy white. 40 cents each.

ACALYPHA, MACAFEANA. Superb Summer bedding plant, with very highly colored bright red leaves. It prefers partial shade. 25 cents.

ABUTILON – Fairy Bells.

Hard wooded greenhouse shrub, blooming almost the entire year; well adapted for house culture, and fine for bedding out in Summer. 15 cents.

BOULE DE NEIGE. A pure white bell-shaped flower, blooming profusely.

DARWINII. Orange scarlet and pink veined flowers; blooms freely in clusters.

MESOPOTANICUM. Of trailing habit, flowers pendant in great profusion.

VEXILLARIUM PICTA. Foliage small, mottled yellow and green; flower scarlet and yellow; of drooping habit.

ARTHUR BELSNAM. Flowers large and of a dark crimson color.

THOMPSONI VARIEGATA. The leaves mottled with yellow.

THOMPSONI PLENA. A fine sport from Thompsoni Variegata; the foliage has retained the same variegation, but the flowers are perfectly double.

GOLDEN FLEECE. A bright yellow; a very profuse bloomer.

ACHYRANTHUS.

Any of the following are suitable to form ribbon lines in contrast with Centaureas, Cineraria, Candidissima, etc. 50 cents per dozen; $4.00 per 100.

AUREA RETICULATA. Foliage beautifully reticulated with bright gold, the stem of bright semi transparent carmine.

EMERSONII. Bright red, lance-leaved.

LINDENII. Rich dark red color, well adapted for either ribbon rows or the edging of flower beds.

ASPARAGUS TENUISSIMUS.

We cannot praise too highly this beautiful new plant. Its fine filmy foliage equals in delicate beauty the Maiden-Hair Ferns. First size, 50 cents; second size, 25 cents; small plants, 10 cents.

AGAPANTHUS UMBELLATUS.

A noble plant belonging to the bulbous-rooted section, with evergreen foliage; the flower stalks grow nearly three feet high, crowned with a head of twenty or thirty blue flowers. 35 cents.

ALYSSUM.

DOUBLE. Very beautiful variety, splendid for cut flowers; fine green foliage, and produces enormous quantities of double pure white, fragrant flowers. 10 cents.

AMARYLLIS.

The Amaryllis are an interesting class of bulbs, desirable for growing in pots, producing showy flowers that are very attractive and handsome.

VITTATA. These magnificent plants are vigorous in their growth, and produce a free supply of flowers, are flaked and striped with the most striking tints, and are justly esteemed the most beautiful of the Amaryllis family. 75 cents each.

JOHNSONI. An elegant pot plant, with crimson flowers five inches in diameter, each petal striped with white. Flower stalk two feet high, with clusters of three to five blooms. 75 cents each.

ACHANIA.

MALVAVISCUS. A greenhouse shrub with scarlet flowers; blooms Summer and Winter; not subject to insects of any kind. One of the most satisfactory house plants every grown. 25 cents.

AGERATUM.

Very useful plants for bedding or borders, flowering continually during the Summer. By cutting back and potting in the Fall they will continue to flower in Winter. 50 cents per dozen; $4.00 per 100.

WHITE CAP. By far the best and most useful variety ever sent out, being a dwarf, compact grower, and bearing profusions of pure white flowers. An exceedingly useful and profitable plant to grow for cut flowers in the Winter, as it blooms freely all Winter.

JOHN DOUGLAS. Azure blue; of compact habit.

MERIDEN GEM. Compact; light blue.

AGAVE – Century Plant.

AMERICANA. Picturesque plant for out-door decoration on the lawn, or for growing in vases. 25 cents to $4.00 each.

AMERICANA VARIEGATA. Similar to above variety, with leaves banded with yellow. These plants stand any amount of heat and drouth, and are therefore admirably adapted for centre plant of vases, baskets, rock work, etc. Small plants, in four-inch pots, 25 to 50 cents; large, one to two feet high, $1.00 to $5.00.

ALTHERNANTHERA.

Plants with beautiful variegated foliage, growing twelve to twenty-four inches in diameter and six inches high. It is used principally for ribbon lines and borders. 5 cents each; 50 cents per dozen; $4.00 per 100.

AMABILIS. Leaves tinted rose.

AUREA. Foliage dark green and golden yellow, the later color predominating.

AUREA NANA. Foliage a bright green, beautifully variegated with yellow.

LATIFOLIA. Broad, smooth, Autumn-tinted leaves.

PARONYCHIOIDES MAJOR. Bronze, tipped red; the brightest and showiest.

SPATULATA. Leaves tinted carmine and green.

ARISTOLOCHIA SIPHO. – Duchman's Pipe.

Very large leaves and brownish flowers of a very singular shape, resembling a pipe; it is a vigorous and rapid growing climber, attaining a height of twenty feet. 25 cents each.

AKEBIA QUINATA.

A climbing plant from Japan, with beautiful cut foliage, having large clusters of chocolate colored flowers, which are very fragrant. Attains the height of twenty feet. 25 cents each.

AMPELOPSIS.

QUINQUEFOLIA. Rapid grower, attaching itself to brick, stone walls or trees; it has beautiful green foliage in Summer, turning to rich crimson in Autumn. 25 cents.

VEITCHII. Miniature variety of Virginia Creeper, which clings to any building, and produces in the greatest profusion dense foliage of glossy pale green, shaded with purple, and which turns brilliantly red in Autumn. Of exceedingly rapid growth, and requires no nailing. It is perfectly hardy. 25 cents each.

ASTERS.

Plants grown from choicest seed. 50 cents per dozen.

ALOYSIA CITRIODORA.

LEMON VERBENA. A favorite garden plant, with delightfully fragrant foliage; fine for bouquets. 10 cents.

ASPIDISTRA LURIDA.

Curious plants, remarkable for producing their flowers under the surface of the earth. The leaves are six inches long, about two inches wide, and of a bright green. They are well adapted for wardian cases, ferneries, etc. 50 cents.

ALLAMANDAS.

The Allamandas are beautiful evergreen climbers, with rich, glossy foliage, and deep yellow flowers, which are very large and showy. It would be difficult to exaggerate the beauty of the Allamandas or their real, permanent value. 50 cents.

BELLIS PERENNIS—Daisies.

Well known ever blooming plants; pink, white and red; double flowering. 10 cents each; 50 cents per dozen.

BILBERGIA SPECIOSA.

Pineapple resembling foliage, with very bright crimson flowers growing out of the heart of the plant; of easy cultivation. 30 cents.

BOUVARDIAS.

These are among the most important plants cultivated for Winter flowers, owing to the yearly increasing variety of color and excellent adaptation for that purpose. They are also effective as bedding plants for the garden, blooming from July until frost. 15 cents each; $1.50 per dozen; small mailing plants, $1.50 per doz.

A. NEUNER. Perfectly double, a pure waxy white, a constant bloomer, and of unsurpassing beauty.

ELEGANS. Salmon scarlet, large and fine; a splendid color.

HUMBOLDTII CORYMBIFLORA. The largest white flowering sort out; flower tubes three inches long; very fragrant.

LEIANTHA. Dazzling scarlet; one of the best, and very profuse.

MAIDEN BLUSH. Bright blush pink flowers, a distinct color; elegant.

THE BRIDE. White, with a very slight tinge of flesh; a really fine sort.

VREELANDII. Finest white; valuable for bouquets; best of all singles.

PRESIDENT CLEVELAND. Extra large fiery-scarlet flowers; vigorous growth.

PRESIDENT GARFIELD. Double pink.

SINGLE FLAVESCENS. Bright canary, very desirable, sweet, reminiscent.

BOCKII. New single pink, producing its flowers in graceful clusters.

CALADIUMS—Fancy Leaved.

We have a fine collection of first-class, distinct. They are never os large as Esculentum, but the brilliant cardinal red, pink, cream and various shades of green that are displayed in the veinings and blotches of the leaves can not be obtained in any other class of plants. 30 cents each for fine, well dried tubers.

CALADIUM ESCULENTUM.

The most striking and distinct ornamental foliage plant in cultivation; desirable for pot or tub culture, and fine for bedding out. With a plentiful supply of water, the leaves may be grown from four to six feet long, and one and one-half feet in breadth. 20 cents each.

COLEUS.

Coleus can be used in numberless ways. Their foliage is superbly colored with bronze, crimson, maroon and gold; sometimes uniform as in Verschaffeltii and Golden Bedder, but usually marked and variegated in the most brilliant manner. Beautiful at all seasons of the year, carrying rich and velvety foliage, and growing rapidly. They are very sensitive to frost, and should not be placed out in the open ground until all danger of cold weather is past. Price 5 cents each, 50 cents per dozen, $4.00 per 100. It will be noticed that we offer these popular bedding plants cheaper than any other firm in the country.

ALINE. Maroon, bright pink centre.

BIZARRE. Deep crimson, centre edged with golden yellow.

CRIMSON BEDDER. Rich dark crimson, with deeper shadings.

CRESCENT. Crimson, violet and gold.

CORSAIR. Rich velvety crimson, bordered with deep maroon.

FIREBRAND. Bright glowing crimson.

GARLAND. Large finely serrated foliage of bright green, purple and crimson.

GOLDEN BEDDER. Clear golden yellow, the best yellow bedder.

JOHN GOODE. Light green, on a yellow ground; a splendid bedder.

JEWEL. Bright crimson, edged yellow.

KENTISH FIRE. Centre clear vermillion, outside green and bronze.

MRS. J. SCHULTZ. Bright golden yellow, marked with scarlet and carmine.

MISS RETTA KIRKPATRICK. Creamy white, centre margined with green.

ONWARD. Large foliage, a deep green, marked and splashed with brown.

PAROQUETTE. Light yellow, distinctly mottled with crimson.

PRISCILLA. Bright green, creamy white markings in centre of leaf.

PROGRESS. Gold, carmine and green.

RED CLOUD. Crimson and dark brown.

ROB ROY. Bright carmine, edged with yellow and green.

ROVER. Crimson, pink and maroon.

SHAH. Violet crimson, marked white.

VERSCHAFFELTII. Velvety crimson.

CANNA.

The Canna is a fine foliage plant, making a good bed alone, but particularly desirable as the centre of a group of foliage plants, for which it is one of the best, growing from three to six feet. Select old sorts, 10 cents.

EHEMANNI. Most distinct of all Cannas on account of its large oval soft green leaves and carmine red flowers, which are produced on long flower stems, each of the smaller branches bearing about twelve flowers. 25 cents each.

NOUTTONI. Is quite distinct from Ehemanni in coloring, being a rich shade of crimson scarlet. The flowers are very large, growing erect instead of drooping. 25 cents each.

CLERODENDRON.

BALFOURI. Very handsome greenhouse climber, with large clusters of crimson-scarlet flowers, each flower encased in a bag-like calyx of pure white. 25 cents.

CATALONIAN JESSAMINE.

JASMINUM GRANDIFLORUM. Beautiful white Jessamine, of exquisite fragrance. 15 to 50 cents.

59

COBÆA.

SCANDENS. Magnificent climber, with large, bell-shaped flowers and elegant leaves and tendrils. It is of rapid growth, and consequently eminently adapted during the Summer for warm situations, where it will produce an abundance of its elegant purple flowers. 20 cents.

CAPE JESAMINE.

GARDENIA FLORIDA. Southern plant of easy cultivation, blooming profusely in Spring and early Summer; flowers a pure white, double; plants very bushy; foliage dark green and glossy. Plants that will bloom, 25 and 50 cents.

JASMINUM REVOLUTUM. A beautiful yellow flowered hardy shrub, and a great favorite in the South.

CINERARIA.

HYBRIDA. These are among the most gorgeous of our greenhouse plants; the colors range through all the shades of blue, violet crimson, pink, maroon and white. They are in bloom only until May. 10, 15, 25 and 30 cents.

COCOLORIUM.

VESPUTALINOUS. Free growing plant of greenhouse culture, suitable for baskets. 20 cents each.

CYCLAMEN PERSICUM.

As an ornamental greenhouse plant it is excelled by few, and its flowers as a variety in the formation of bouquets and baskets of cut flowers in Winter are valuable. 10 to 25 cents.

CACTUS.

Of these plants we have a fine collection. The Cactus family is interesting on account of the curious leafless growth of the plants and the beauty of the flowers, the Lobster Cactus, especially, being a great favorite.

EPIPHYLLUM TRUNCATUM. Lobster Cactus. Winter blooming. 25 cents.

CEREUS GRANDIFLORUS. The Night Blooming Cereus. 25 cents.

CENTAUREA.

GYMNOCARPA. Dusty Miller; attains a diameter of two feet, forming a graceful round bush of silver gray, for which nothing is so well to contrast in ribbon lines with dark foliaged plants. 50 cents per dozen.

CESTRUM PARQUI.

The Night-Blooming Jessamine. This well known and very highly prized plant, blooming nearly all the year round, is a native of Chili. It is an excellent garden plant, growing rapidly, the foliage long and of a deep green color: producing its richly fragrant flowers at every joint, sweet only at night. It is well adapted to house and window culture. 10 cents each.

CISSUS DISCOLOR.

A well known climber, with leaves beautifully shaded dark green, purple and white, the upper surface of the leaf having a rich velvet like appearance. 15 to 30 cents each.

CHRYSANTHEMUM FRUTESCENS

This is the Paris Daisy now so fashionable and in such demand during the Winter. The flowers much resemble our common field Daisy; almost constant in bloom. 10 cents each.

CROTONS.

The Crotons are among the finest decorative foliage plants known. The leaves of all are more or less veined and margined, sometimes entirely variegated with shades of yellow, orange and crimson. Some have long, narrow leaves, arching gracefully, in fountain-fashion; others broad and short, oak-leaved. Some recurved very much, others are twisted, cork-screw like. Crotons love heat, sunshine and moisture. They make beautiful bedding plants in the heat of Summer. 50 cents each; small plants, 25 cents.

AUGUSTIFOLIA. Narrow leaved, yellow and red.

AUREA. Small foliage, yellow mottled on dark green ground.

CORNUTUM. Distinct and very compact growing variety, blotched and spotted yellow.

DISCOLOR. Light green leaves, a claret color on reverse side of leaf.

EARL OF DERBY. Leaves highly colored with bright yellow.

INTERRUPTUM. This is one of the finest and most elegant of the many Crotons. It is a finely marked variety, with dark red variegation.

UNDULATUM. Foliage metallic green, spotted crimson, pink and yellow, the edges of the leaves being beautifully undulated and wavy.

VEITCHII. Leaves waxy green, marked with yellow, changing to rose, scarlet purple.

VARIEGATUM. Leaves a dark green, striped and spotted golden yellow.

VOLUTUM. Very distinct and beautiful form, the great peculiarity of which consists in the leaves being rolled up from the end in a volute, after the manner of the curving of a ram's horn.

YOUNGII. Leaves eighteen inches long, very distinct, noble and graceful habit, surface dark green, marked with creamy yellow and bright rosy red.

EUPHORBIAS.

Plants of great value for Winter blooming and making splendid pot plants; they are sure to bloom with regularity, are easily cared for, and do not suffer much from a moderate amount of neglect or abuse. 25 cents each.

CYPERUS.

ALTERNIFOLIUS. Grass-like plant, rearing up its stems to the height of about two feet, surmounted at the top by a cluster or whorl of leaves, diverging horizontally, giving the plant a very curious appearance. A splendid plant for the centre of baskets or wardian cases, or as a water plant. 10 cents each.

COCOLOBA.

PLATYCLADA. Plant of singular and interesting growth, stem and branches growing to flat, broad points. It is well suited for vases and rustic work. 10 cents each.

CUPHEA—Cigar Plant.

PLATYCENTRA. Tube of scarlet, tip white and black, very free bloomer. A good basket plant, and also an excellent plant for the house in Winter. 10 cents each.

DRACÆNA—Dragon Tree.

One of the most desirable of our ornamental foliage plants for decoration either in or out doors, as it does not appear to suffer under the dry atmosphere of rooms. In a partially shaded situation it stands remarkably well during the Summer out of doors. 25, 50 and 75 cents and $1.00 each.

BAPTISTI. Green, creamy white flakes flushed with rose.

IMPERIALIS. Strong growing variety, large deep rose and creamy white foliage.

TERMINALIS. Rich crimson foliage marked with pink and white.

YOUNGII. Light green changing to copper color.

INVIVISA. Long foliage, green, graceful.

EUCHARIS.

AMAZONICA. The beautiful American Lily. A bulbous rooted plant, with very broad Lily-like leaves and pure white flowers about four inches in diameter, borne in heads of four or five, and deliciously fragrant. Give them an abundance of water when growing and blooming. Fine bulbs, 50 cents.

ECHEVARAS.

A genus of succulent plants, natives of Mexico. They are of rich appearance, and well suited for rock work.

SANGUINEA. Narrow pointed leaves; color a deep red. 15 cents.

SPLENDENS. The plant is a perfect mass of thorns, and anything but handsome, the numerous flowers, however, being bright and beautiful. They are brilliant scarlet, and borne in clusters of six or seven. Almost always in bloom, and requires but little water. 25 cents each.

SECUNDA GLAUCA. Dwarf sort, resembling the house leek; glaucus green. They bloom all Summer; an excellent plant for borders or rock work. 10 cents each; $1.00 per dozen; $6.00 per 100.

FREEZIA REFRACTA ALBA.

A bulbous rooted plant of the easiest cultivation. The flower is pure white, spotted with lemon yellow. In shape it is like a miniature Gladiolus, only more extended and deliciously sweet. Will bloom best planted out during Summer. 10 cents each.

FEVERFEW.

LITTLE GEM. The finest double white raised, blooming very freely and being more dwarf, with larger and more double flowers than the old variety; a first-class plant, that everyone should have. 10 cents each.

DOUBLE WHITE. Very free blooming, double, Daisy-like flower. Very useful for Summer bouquets.

FICAS.

ELASTICA. India Rubber Tree. One of the best plants for table or parlor decoration. Its thick leathery leaves enable it to stand excessive heat and dryness, while its deep glossy green color always presents a cheerful aspect. The plants we offer are in fine order, and are of a size to be useful immediately. 75 cents to $1.50 each.

FORGET-ME-NOT.

MYOSOTIS PALUSTRIS. Requires no description, its clustered flowers of beautiful blue having had a place in romance and literature since romance and literature began. 10 cents.

FERNS.

These very beautiful plants are now very generally cultivated, their great diversity of gracefulness of foliage making them much valued as plants for vases, baskets, or rock work, or as specimen plants for parlor and conservatory. 15 cents.

GREVILLEA ROBUSTA.

The Australian "Silk Oak." A splendid ferny-leaved tree, evergreen, and especially adapted as a shade tree. Thousands are being annually planted. Also used by florists for decorating apartments, etc. A magnificent pot plant. 25 cents.

GLADIOLI.

Among bulbous flowers the Gladiolus deserves first place in popular favor. Our collection is very fine, a good assortment of colors, red, pink, striped and many shades of light colors. By express, 75 cents per dozen; by mail, $1.00 per dozen.

HYDRANGEAS.

HORTENSIA. The well known garden variety. Has immense heads of pink flowers, which hang on for months. 15 cents.

OTAKSA. Heads large, bright rosy pink, contrasting beautifully with other sorts. Of low bushy growth. 10 cents.

THOMAS HOGG. Immense truss, at first tinged with green, then turning a pure white. 15 cents.

GLOXINIA.

Gloxinias are among the handsomest of our Summer blooming greenhouse plants. Bulbs should be started in the Spring, in a warm place. They require partial shade and a liberal supply of water when growing. After blooming, water should be withheld, and the bulbs remain dry during the Winter. 50 cents each; small plants, 15 cents each.

HIBISCUS.

A beautiful class of greenhouse shrubs, with handsome glossy foliage, and large, showy flowers, often measuring over four inches in diameter. They succeed admirably bedded out during the Summer. 15 cents each.

BRILLIANTISSIMA. Single flowers, of the richest crimson scarlet; dark crimson at base of petals; very large and showy.

DENISONII ROSEA. Large, single flowers, a clear transparent rose.

GRANDIFLORA. Enormous rosy crimson, single flowers, produced in abundance.

KERMESINUS. Enormous and very double rich carmine crimson.

MINIATUS SEMI-PLENUS. Immense semi-double flowers, dark vermillion scarlet.

ZEBRINUS. Outer petals scarlet, edged yellow, variegated yellow and scarlet.

HOYA—Wax Plant.

CARNOSA. Star-shaped, waxy flowers, in clusters; beautiful, thick, glossy, evergreen leaves. Excellent for house decoration. 25 cents.

HELIOTROPE.

A great favorite on account of the delicate fragrance of its flowers; a constant bloomer when planted out in a sunny, warm place, the colors varying from nearly white to dark purple. By express, our selection, 25 cents; by mail, 20 for $1.00. Purchaser's selection, 10 cents each, 75 cents per dozen.

PRESIDENT GARFIELD. A gem of the first water, fine deep blue, very floriferous.

CHIEFTAIN. Rich shade of violet, the best Winter bloomer.

JERSEY BEAUTY. The finest blue variety; best for pot culture; dwarf.

Mrs. David Wood. Flowers semi-double, in large heads, fragrant, early, constant bloomer, light blue.

Albert Delaux. Bright golden yellow foliage, marked with delicate green, and lavender flowers.

White Lady. Strong growing and free branching, very profuse in bloom, large and of the purest white.

IMPATIENS.

Sultani. A new plant of the same order as the well known Balsams, but differing widely from them in the habit of blooming. The flowers are borne in clusters or masses around the head of the plant. 15 cents each.

JASMINUM.

Grandiflorum. The Catalonian Jessamine. Winter-flowering plants, blooming without intermission from October to May. The flowers are pure white, and most deliciously fragrant. 15, 25 and 50 cents each.

Grand Duke. Flowers are double, white like miniature Rose, and deliciously fragrant. 75 cents each.

LANTANAS.

Plants much used for bedding and pot culture. They are strong growing and constant bloomers. 10 cents each; $1.00 per dozen.

Araxtiaca. Beautiful orange.

Jacob Schultz. Red, changing to crimson.

Purpurea. Good purple.

Rosa Myxni. White and rose.

LIBONIAS.

Floribunda. Long flowers, shaded from orange scarlet at the base to deep yellow at the mouth. 10 cents.

Penrhosiensis. Winter blooming plant of neat and pretty habit. 10 cents.

NASTURTIUM.

Empress of India. The plant is of a very dwarf habit, with dark tinted foliage, while the flowers are of the most brilliant crimson, so freely produced that no other annual in cultivation can approach it in effectiveness. 10 cents.

OXALIS.

These plants are of the easiest possible culture, and are fine for baskets, vases, etc.

Lutea. Large clear yellow flowers, in the greatest profusion, 15 cents.

Rubra. Flowers bright red.

White. Color white, flowers profusely Summer and Winter. 10 cents.

OLEANDER.

Double Pink. The oldest and finest of all varieties in cultivation; flowers double and rose colored. 20 cents.

Lilian Henderson. New double, and one of the best yet introduced. 50 cents.

PASSIFLORA—Passion Flower.

Will bloom a long time in the house if grown in a large pot or tub and removed before frost. 50 cents each.

Quadrangularis Folia Variegata. This is a magnificent novelty. The foliage is beautiful in itself, of a deep olive green, blotched and dotted with rich golden yellow. The flowers are very large and sweet scented. Color purple inside of petals, light green on the outside; the centre of the flower is of many colors.

PRIMULA OBCONINA.

This is undoubtedly one of the most useful flowering plants grown. The seedlings will begin to bloom in May or June and continue to bloom during the whole year. 20 cents each; $2.00 per dozen.

MUSA ENSETE.

The noblest of all plants is this great Abyssinian Banana. The leaves are magnificent, long, broad and of a beautiful green, with a broad crimson midrib. The plant grows luxuriantly from eight to twelve feet high. During the hot Summer, when planted out, it grows rapidly and attains gigantic proportions, producing a tropical effect on the lawn, terrace or flower garden. We offer a fine lot of these plants, at $1.00, $1.50 $2.00 and $3.00 each. A few extra strong plants, $5.00 each.

MAHERNIA ODORATA.

A profuse Winter-blooming plant, with golden yellow flowers, that emit a strong honeyed fragrance. 10 cents each.

PILEA.

Arborea, The Artillery plant. Pretty little plant of drooping habit, resembling the Fern; fine basket plant. 15 cts.

PLUMBAGO.

The Plumbagos are desirable on account of their beautiful shades of blue, a color by no means too common among the flowering plants.

Capensis. Very bright plants, large heads of light blue flowers. 15 cents each.

PRIMROSE—Chinese.

Few house plants afford more genuine satisfaction than this. It requires to be kept cool, a north window suiting it best. Primroses are at present all in bloom. 20 cents each; $2.00 per dozen.

PETUNIAS—Double.

Few plants have been so much improved as the Petunias. The double flowers are of much greater size than the largest of the singles, and are very richly colored. They flower freely, and continue often even after hard frost. They make splendid pot plants for early Spring blooming. 15 cents each.

PANSIES—New Double Giant.

This class of plants cannot be overestimated. The gigantic size of the flowers, luxuriant growth, profusion of bloom, and exquisite blending of gay and fantastic color, is utterly indescribable. The colors are truly wonderful, including many different shades and combinations. We believe that our fine new Giant Pansies are the finest strain ever offered. 5 cents each. 25 for $1.00, $3.50 per 100.

Trimardeau. Flowers of immense size and splendid shape, distinctly marked with three blotches, and are from three to four inches across.

Bugnot's New Spotted. Somewhat smaller than Trimardeau, but of even more exquisite markings and richer colors.

Faust. King of the Blacks. This is the darkest Pansy known, a perfect beauty.

Cassier's Odier. Flowers are of immense size, three or five spotted on backgrounds of very rich colors.

New German. The new German Mixed Pansies are in almost endless variety of charming shades and color, united with extra large size of flowers.

RUSSELIA JUNCEA.

Has long, very graceful, rush-like foliage, the drooping tips of which bear tubular, light scarlet blossoms in showers; there is nothing so beautiful for large vases. A handsome house plant. 25 cents.

62

POINSETTIA PULCHERRIMA.

A new double Poinsettia. A very brilliant scarlet, tinted with orange color, a dazzling color. The head grows on a specimen plant fourteen inches in diameter by ten inches in depth, giving it the appearance of a cone of fire. 25 cents each.

RHINCOSPERMUM

JASMINOIDES. Greenhouse climber, with white Jessamine-like flowers, which are produced in great clusters in the Spring months, and have a delicious fragrance. 25 cents.

SALVIA—Flowering Sage.

This plant is indispensable in the garden in Autumn. They may be planted in masses or scattered among the shrubbery, in either way their gorgeous effect is well displayed, 10 cents each, except where noted. Where the selection is left to us, 20 for $1.00.

MRS. EDWARD MITCHELL. Crimson purple.
PATENS. Exquisite blue.
RUTILANS. Magenta, apple fragrance.
SPLENDENS. Brilliant scarlet, beautiful.
SPLENDENS ALBA. White flowered.

TRADESCANTIA

ZEBRINA. Wandering Jew. Leaves striped a silvery white. $1.00 per dozen.

SMILAX.

A climbing plant, unsurpassed in the graceful beauty of its foliage. Its peculiar wavy formation renders it one of the most valuable plants for bouquets, wreaths, festoons and decorations. 15 cents each.

TUBEROSE.

PEARL. New double. Flowers of large size, imbricated like a Rose, dwarf habit, growing only from eighteen inches to two feet high. 10 cents each; 50 cents dozen; $3.00 per 100; by express, 30 for $1.00.

VIOLETS.

It is one of the leading florists' flowers for bouquets and cut flowers. All the varieties should have a slight protection of leaves during the Winter. A better plan to insure early Spring flowers is, to plant in cold frames in the Fall. They thrive best in a shady situation, in rich, deep soil. 10 cents each.

BLUE NEAPOLITAN. Double light lavender blue, very profuse bloomer.
MARIE LOUISE. Double, darker than the above, and larger in size.
CZAR. Single, rich bluish purple, large.
SCHOENBRUN. Single, dark blue, profuse.
SINGLE WHITE. White blooming.
VICTORIA REGINA. Largest single, purple.
SWANLEY WHITE. Pure white, large size.

VINCA—Periwinkle.

Best blooming plant for bedding out, being constantly in bloom from June until frost, bearing the hot sun and frequent drouth well, and is excellent for the South. We have a good stock. 10 cents each; 50 cents per dozen; $1.00 per 100.

ALBA. Pure white, hundreds on a plant.
ROSEA ALBA. Pure white, dark rose eye.
ROSEA. Dark rose pink.

WISTERIA SINENSIS.

One of the most hardy climbing plants, and when once established, of very rapid growth, covering the entire side of a house in a few years, presenting a magnificent appearance when in full bloom, with its thousands of rich clusters in pendulous racemes of delicate violet blue blossoms, richly perfumed. 50 cents.

Hardy Herbaceous Shrubs.

Deutzia Gracilis

Graceful white blooms, produced all the Spring in large quantities; are dwarf and bushy. 25 cents each.

Lychnis Calceodonica.

A beautiful Summer-flowering plant entirely hardy. Flowers in June. 10 cents.

Campanula Carpatica.

The old Canterbury bell. A beautiful cup-shaped garden flower and a great favorite everywhere. 10 cents each.

Apiosa Tuberosa.

Hardy bulbous rooted vine, known also as the Manetta vine, makes a nice shade and flowers freely. 20 cents.

Dielytra Spectabilis.

Hardy ornamental flowering plant, also known as the Bleeding Heart; a valuable garden plant. 20 cents.

Erianthus Ravennæ.

Perfectly hardy, the foliage forms graceful clumps three to four feet high, above which arise numerous spikes five to six feet, bearing plumy flowers. 50 cents each.

Azalea Americana.

Hardy Azaleas. Are deciduous, flowering in May. Light straw-colored blossoms, very beautiful. 50 cents each.

Iris Susiana—The Mourning Bride.

The groundwork of the flower is a silvery gray, shaded and lined with very dark chocolate and black. 25 cents each.

Ixias.

These are amongst the most graceful and beautiful of half-hardy bulbs. 25 cents.

Hollyhocks.

Superb double kinds. The Hollyhock is becoming a very popular Summer-flowering plant, and when planted in rich soil and sunny position it is a very impressive and stately plant. We offer strong, one year old plants, at $1.00 per dozen. Nice young plants that will bloom this year, 50 cents per dozen.

Astibile Japonica.

Incomparably the most beautiful of all hardy herbaceous plants, growing about two feet high, in compact shape, with handsome foliage, from above which rise its panicles of small, feathery, white blossoms. 25 cents each.

Anemoue—Wind Flower.

A very pleasing hardy perennial, bulbous rooted plant, easily grown from seed, producing very large flowers early in the Spring, in a sunny situation. 10 cents each.

Dianthus—Pinks.

A magnificent genus, embracing some of the most popular flowers in cultivation. 10 cents each.

Iris—Fleur-de-Lis.

The Iris is a very extensive and beautiful family, commonly known as the Flowering Flag. 25 cents each.

Wiegelias.

Beautiful shrubs that bloom in June and July. The flowers are produced so so great profusion as to almost hide the foliage. 25 cents each.

Viburnum Plicatum—Jap. Snowball.

A beautiful shrub of moderate upright growth, with crinkled or plicated rich green leaves. The flowers are white, and larger and more solid than those of the common Snowball. 25 and 50 cents.

63

Pæonias.
Pæonias. like other meritorious plants,
have always admirers. 30 cents each.
Philadelphus—Mock Orange.
Coronarius. Medium sized shrub, bear-
ing an abundance of white, sweet-scented
flowers; last of May. 25 cents.
Deutzia.
Crenata. Height, 2 to 3 feet, regular and
compact form, bushy, flowers pure white,
blooms profusely, hardy. 25 cents.
Crape Myrtle.
Pink. Fringed pink blossoms.
Crimson. Deep crimson. 10 cents.
Hydrangea Paniculata Grandiflora.
One of the finest hardy shrubs in cultivation; the flowers are formed in large
white panicles of trusses, nine inches to
one foot in length. 25 to 50 cents each.
Tamarix Gallica.
The pink flowers of the Tamarisk, borne
all along its slender branches, and its delicate feathery foliage give it a character no
other shrub possesses. 50 cents each.
Phlox.
Our collection embraces the best of the
old varieties and the new French ones of
recent introddetion, which are very fine,
distinct, pure colors, many of them beautifully shaded and marked with distinct
clear, light eyes. 15 cents; $1.50 per dozen

Vegetable Plants.

Asparagus.
Connover's Collossal. Two year old
roots, $1.00 per 100; $6.00 per 1,000.
Rhubarb.
Or Pie Plant. $1.50 per dozen.
Tomato Plants.
We grow all the newest and best, Mika-
do, Advance, Thoburns, Volunteer, New
Jersey, Dwarf Champion and many others.
We grow them in small pots, and they
can be shipped with the greatest safety,
and not disturbing the roots. By express
50 cents per dozen; $4.00 per 100. Ready
after February 1st.
Egg.
25 cents per dozen.
Pepper.
25 cents per dozen.
Sweet Potato.
30 cents per 100; $2.50 per 1,000.
Cabbage.
We have a handsome lot of about 100,000
Cabbage plants, wintered in cold frames,
of the following varieties: Landreth's Earliest, Winning Star, Early Flat Dutch,
Early Drumhead, Select Jersey, Wakefield
and others. Price per 100, 50 cents; $4.00
per 1,000; in lots of more than 1,000 a special
rate will be given. The plants are fine
and can be shipped any time after January
1st. Young plants from seed bed after
April 1st. $2.00 per 1,000.

Grapes, Raspberries, Etc.

Grapes.
Concord. Black; best for general cultiva-
tion. Two years, 15 cents each.
Ives' Seedling. Dark purple. Two years,
15 cents each.
Moore's Early. Large black, excellent.
Two years, 25 cents each.

Pocklington. Golden yellow. Two years,
30 cents each.
Hartford Prolific. Black, fine and early.
Two years, 15 cents each.
Prentiss. Greenish white. Two years, 30
cents each.
Raspberries.
Price, 50 cents per dozen; $2.50 per 100.
Turner. Very hardy, which character
makes it the favorite in the South.
Gregg. One of the best and largest.
Hansel. One of the earliest, bright scarlet.
Cuthbertii. Rich and luscious, crimson.
Gooseberries.
Downing. Very large, handsome green
of splendid quality for both cooking and
table use. 20 cents; $2.00 per dozen.
Houghton Seedling. Small to medium;
pale red, roundish oval, sweet, tender,
very good. 15 cents; $1.50 per dozen.
Currants.
Large, two years, 15 cents; $1.50 per dozen.
Red Dutch. Old reliable sort.
White Grape. The best white.
Black Naples. Good old variety.
Blackberries.
Early Harvest. Very early.
Kittatiny. Large berry.
Snyder. Hardiest of all. 50 cents per
dozen; $2.50 per 100.
Strawberries.
We grow the following varieties in quan-
tity at 75 cents per 100, $5.00 per 1,000; Chas.
Downing, Kentucky, Cumberland Tri-
umph, Sharpless, May King, Wilson's
Albany, Jersey Queen. We have found
the following two varieties to be especially
suited to this climate, and highly recommend them. $1.00 per 100; $7.00 per 1,000.
The "Henderson." The fruit is of the
largest size, early, and immensely productive, but its excelling merit is its exquisite flavor.
Hoffman's Seedlings. This is a variety
raised in North Carolina, and it is particularly adapted to the South. It is
early, and stands the dry Summers better
than any other variety we grow.
Strawberry plants can only be sent by
express. At this rate, if wanted by mail,
add 10 cents per 25, or 30 cents per 100 for
postage.

Gold Fish and Globes.

There is nothing more attractive in a
room than Gold Fish, and there is but
little trouble in keeping them. Change
the water two or three times a week, river
or cistern water will do, and clean the globe
once a week. Feed the fish wafer crackers
a little at a time. They can be sent any
distance by express in tin cans at buyer's
risk. Large finely colored fish, 50 cents
each. Cans for shipping fish, 15 cents each.
We have a nice lot of fish globes, large
enough to hold two gallons of water, and
amply large for a pair of our largest fish,
that can be packed securely and sent by
freight or express. Price $2.00 each.

Lawn Grass Mixture.

Finest prepared, per bushel, $3.00; per
peck, $1.00.

www.ingramcontent.com/pod-product-compliance
Lightning Source LLC
Chambersburg PA
CBHW022008190326
41519CB00010B/1441